Metrology for Engineers

The development of engineering production in the past has been inseparable from that of metrology; and it is certain that industry's increasing demands for stringent design requirements will result in even closer attention being paid to the science of measurement in the future.

Accordingly, *Metrology for Engineers* has been written to meet the needs of students preparing for examinations in metrology or in the subjects of which metrology forms a part. It is specifically designed to prepare students for the following examinations: the Higher Diploma and Certificate examinations in Engineering, the Membership examinations of the Council of Engineering Institutions, and the Institute of Quality Assurance. *Metrology for Engineers* also covers adequately the standard unit (TEC U79/637) for the Higher Certificate and Diploma in Mechanical and Production Engineering (Programme A5 of the Technician Education Council).

The work has been set out to ensure the greatest possible accessibility of the material, and it is comprehensively illustrated.

ALSO BY C. R. SHOTBOLT:

Workshop Technology for Mechanical Engineering Technicians, Books 1 and 2

Technican Workshop Processess and Materials I

Technician Manufacturing Technology 2 and 3

Metrology for Engineers

FOURTH EDITION

by J. F. W. GALYER

Formerly Principal Lecturer in Production Engineering at Luton College of Technology

and C. R. SHOTBOLT

C.Eng., F.I.Prod.E., M.I.Mech.E., M.I.Q.A., M.A.S.Q.C.

Principal Lecturer in Production Engineering at Luton College of Technology

CASSELL

Cassell Publishers Limited
Artillery House, Artillery Row, London SW1P 1RT

First edition April 1964
Second edition (revised) January 1968
Third (revised) edition August 1969
Third edition, second impression February 1971
Third edition, third impression September 1972
Third edition, fourth impression December 1974
Third edition, fifth impression March 1977
Third edition, sixth impression February 1979
Fourth (revised) edition July 1980
Fourth edition, second impression July 1981
Fourth edition, third impression August 1983
Fourth edition, fourth impression August 1985
Fourth edition, fifth impression September 1986
Fourth edition, sixth impression January 1988
Fourth edition, seventh impression March 1989

ISBN 0 304 30612 6

Printed in Hong Kong by Colorcraft Ltd

Preface to the First Edition

THE purpose of this book is that it should meet the needs of students preparing for examinations in the subject of metrology, or in that group of subjects which metrology forms part. It is designed to be appropriate reading in preparation for the following examinations in particular: the Higher National Diploma and Certificate examinations in Mechanical and Production Engineering, and the Associate Membership examinations of the Institution of Mechanical Engineers, the Institution of Production Engineers, and the Institute of Quality Assurance. It will also prove useful to those students preparing for the final years of the Mechanical Engineering Technicians' Courses of the City and Guilds of London Institute.

A knowledge of metrology, and practice in its application, is of increasing importance in industry. This is true also of those techniques which are allied to metrology and do not strictly form part of it. It is for this reason that a chapter dealing with statistical quality control has been included. It is hoped that not only will this be of benefit to college students, but also to those in industry who require a knowledge of the fundamentals of this technique.

The development of engineering production in the past has been inseparable from that of metrology. It is certain that the increasingly exacting demands of industry for mechanisms and assemblies generally, the functioning of which must meet stringent design requirements, will result in even closer attention being paid to the science of measurement.

The writers would commend to the student the words of Lord Kelvin, which were used by Dr. H. Barrell in the opening remarks of his Sir Alfred Herbert Paper, 'The Bases of Measurement', presented to the Institution of Production Engineers in 1957: 'I often say that when you can measure what you are speaking about and express it in numbers you know something about it; but when you cannot measure it, when you cannot express it in numbers, your knowledge is of a meagre and unsatisfactory kind.'

This statement presents a powerful case for the subject of metrology to be part of the studies required of an engineer. It also points the way to the fact that if we are to further improve the control of the processes of manufacture, we must continuously develop our means of measurement. This book will have achieved its purpose if it offers to students the basic knowledge upon which the necessary further development may be built.

We wish to express our gratitude to the several firms who have been kind enough to assist in the preparation of the book by granting permission to publish various illustrations, all of which are acknowledged in the text.

Luton. December 1963   J. F. W. G.
                                                                 C. R. S.

Preface to the Fourth Edition

SINCE the third edition of this work was published to cater for the introduction of SI units there have been a number of changes in the British Standards Specifications of which mention is made. This edition is largely, therefore, an updating of the text to cater for these changes, particularly in the field of limit gauges, where B.S. 969 has been superseded by B.S. 4500 Part 2 and in the field of screw thread gauges.

At the same time the opportunity has been taken to introduce a section on the measurement of roundness. For convenience this has been included in chapter 9 which deals with surface finish, but it must be emphasized that one is concerned with geometry and the other with texture. At the time of the revision great changes are taking place in technical education with the phasing out of the Higher National Certificate and City & Guilds of London Institute course 255. Whatever changes are made there will still be a need for the ability to handle precise measuring equipment and it is hoped that this book will be of interest to students taking specialized level 4 and 5 units in the new courses validated by the Technician Education Council (TEC) as well as to those engaged in the HND and CEI part 2 course, many of whom have been kind enough to tell us in the past how useful the work was.

Luton 1980

J. F. W. G.
C. R. S.

Contents

CHAPTER 1

Errors in Measurement

1.1 SCOPE

ALL engineers, regardless of the branch of the profession to which they belong, are constantly faced with the problem of measurement. It may be of time, mass, force, temperature, the flow of an electric current, length, angle, and so on; or it may be of the effects of some of these in combination. Almost invariably, the results of such measurements will determine the course of action the engineer takes thereafter. Thus the results obtained by measurements provide information upon which decisions are made.

All such measurements form part of the science of metrology. The mechanical and production engineer are, however, especially concerned with the measurement of length and angle. Of these, length is of fundamental importance since angular measurement may be carried out by the appropriate use of linear measurements in combination.

Thus the purpose of any measurement is to provide a service to enable a decision to be made. The service will not be complete unless the measurement is made to an acceptable degree of accuracy, but it must be realized that no measurement is exact. It is therefore necessary to state not only the measured dimension, but also the *accuracy of determination* to which the measurement has been made. As far as possible the errors inherent in the method of measurement used should be kept to a minimum, and having minimized the error, its probable magnitude, or accuracy of determination, should be stated.

It follows that it is not enough to state that the nominal size of a gauge block is, for example, 30 mm. It is also necessary to state:

(*a*) The measured error in the block, e.g. -0.0002 mm.

(*b*) The accuracy of determination, e.g. ± 0.0004 mm.

The user, having this information, may now avail himself of it if necessary. If the gauge block is used to set the datum for a vernier height gauge which can only be read to 0.02 mm, then the gauge block errors are negligible and can be ignored. If on the other hand, it is used to set up a comparator whose scale divisions represent 0.001 mm, then the measured error is significant and must be considered, and the accuracy of determination of the gauge block must be incorporated in the accuracy of determination of the comparison being made.

1

It must be pointed out here that in this chapter reference will be made to subsequent work throughout the book.

1.2 TYPES OF ERROR

Generally the errors incurred in any measurement can be considered to be of two distinct types, those which should not occur and can be eliminated by careful work and attention to detail, and those which are inherent in the measuring process.

1.21 Errors Which Can Be Largely Eliminated

1.211 Calamitous or Catastrophic Errors

These are errors of large magnitude having two fundamental causes:

(*a*) *Misreading an instrument*. A micrometer is misread as 6·28 mm or 5·78 mm instead of the correct reading of 5·28 mm.

(*b*) *Arithmetic errors*. These are usually errors of addition. A simple check is to make the calculation twice using different methods, e.g. add a column of figures twice, first upwards then downwards, to ensure that the two results coincide.

In most cases such errors give a result so different from that expected that it is obvious when an error has occurred, and the measurement is repeated and the error detected. This may not always be so, however, and such errors can only be avoided by care and attention to detail.

1.212 Alignment Errors

This type of error occurs when the measuring instrument is misaligned relative to the workpiece. It usually results in the measured dimension M being related to the actual dimension D by one of the trigonometrical ratios. Hence

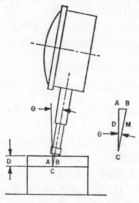

Fig. 1.1. Cosine error due to misalignment of measuring instrument.

such errors are known as trigonometrical or cosine errors. A simple example is shown in Fig. 1.1, where a dial gauge is inclined at angle θ to the required line of measurement. It can be seen that $D = M \cos \theta$.

Another form of this error is parallax, where the line of sight is not normal to the instrument scale. This problem is discussed further on page 50.

1.213 Errors Due to Ambient Conditions

Most measurements are affected to a greater or lesser extent by the environment in which they are carried out. The most important condition is the temperature, both of the workpiece and of its surroundings. The international standard temperature of measurement is 20°C (68°F) and the ambient temperature should be maintained at this level. However carefully this is controlled, it is to no avail if the temperature of the workpiece is allowed to vary. Handling a gauge changes its temperature, so it should be handled as little as possible, and having been handled, allowed to stabilize. Where measurements are being made to a high order of accuracy a time of 20 minutes per 25 mm length of gauge is recommended. During a measurement it is best if all of the components used are left standing on a cast iron surface plate rather than a plastic or wooden bench top. The cast iron, being a good conductor, acts as a heat sink and dissipates temperature differentials more rapidly.

There are two situations to be considered when the effects of temperature are to be discussed:

(a) *Direct measurement.* Consider a gauge block being measured directly by interferometry (see Chapter 2). Here the effect of using a non-standard temperature produces a proportional error:

$$\text{Error} = l\alpha(t - t_s)$$

where
l = nominal length
α = coefficient of expansion
$(t - t_s)$ = deviation from standard temperature

(b) *Comparative measurement.* If we consider two gauges whose expansion coefficients are respectively α_1 and α_2, then the error due to a non-standard temperature will be

$$\text{Error} = l(\alpha_1 - \alpha_2)(t - t_s)$$

As the expansion coefficients are small numbers, the error will be very small as long as *both parts are at the same temperature.* Thus in comparative measurements it is important that all components in the measuring system are at the *same* temperature rather than necessarily at standard temperature.

Other ambient conditions may affect the result of a measurement. If a gauge block is being measured by interferometry, then relative humidity, atmospheric pressure and carbon dioxide content of the air affect the refractive index of the atmosphere. These conditions should all be recorded during the test and the necessary corrections made.

1.214 Errors Due to Elastic Deformation

Any elastic body subject to a load will undergo elastic deformation. The magnitude of the deformation will depend upon the magnitude of the load, the area of contact and the mechanical properties of the materials in contact. It is therefore necessary to ensure that the measuring loads are the same in comparative measurement.

In most instruments used in fine measurement, comparators, bench micrometers, etc., the measuring pressure is reasonably constant, and it follows that the greatest difficulty is due to different types of contact when first setting an instrument to a gauge and then taking a reading on the work under test. A striking example of this is in the measurement of the simple effective diameter of a screw thread where the setting master requires two-point contacts and the thread has four-point contacts in the vee form. Tables of corrections are published* and may be used if the required accuracy warrants such correction.

If a comparison is to be made to a high order of accuracy between components of different radius and from materials whose elastic properties differ, notably the elastic modulus E and Poisson's ratio ν, then correction can be made for the difference in elastic deformation which will occur when the measuring stylus is brought into contact with the setting gauge and the workpiece.

The expression below, derived from the work of S. Timoshenko,† describes the deformation, i.e. the change in centre distance between a spherical stylus and a spherical surface when they are brought together under a pressure W.

$$\text{Total deformation } \delta = 1 \cdot 774 \; W^{2/3}(k_1+k_2)^{2/3}\left(\frac{1}{R_1}+\frac{1}{R_2}\right)^{1/3}$$

in which $R_1 = $ radius of gauge $\qquad R_2 = $ radius of measuring stylus

$$k_1 = \frac{1-\nu_1{}^2}{E_1} \qquad\qquad k_2 = \frac{1-\nu_2{}^2}{E_2}$$

If the comparative measurement consists of setting the comparator to a gauge A and taking a reading on gauge B, then the error incurred will be $(\delta_A - \delta_B)$, appropriate values for k_1, k_2, R_1 and R_2 being inserted in the expression for δ for both situations, A and B.

Another form of elastic deformation is that which occurs when a body sags under its own weight. This problem was considered by Sir G. B. Airy, who showed that the positions of the supports can be arranged to give a minimum error. Two conditions are considered, both shown in Fig. 1.2, one where the slope at the ends of the bar is zero and the other where the deflection at the ends is equal to the deflection at the centre. In the case of line standards the bar is made of 'H' section with the scale engraved on a surface in the plane of the neutral axis. Thus the elastic deformation due to sag has the minimum effect on the length of the scale divisions.

---

* *Notes on Applied Science* No. 1: *Gauging and Measuring Screw Threads*. (N.P.L.) H.M.S.O.
† S. Timoshenko, *Theory of Elasticity*. McGraw-Hill.

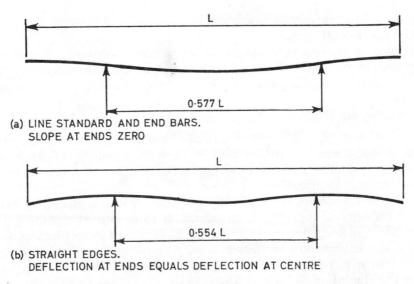

(a) LINE STANDARD AND END BARS.
SLOPE AT ENDS ZERO

(b) STRAIGHT EDGES.
DEFLECTION AT ENDS EQUALS DEFLECTION AT CENTRE

Fig. 1.2. Support positions for different conditions of measurement.

1.22 Errors Which Cannot Be Eliminated

No measurement can be made to give an exact dimension. Fundamentally this is because eventually the numerical value recorded depends upon the human eye reading a scale. The reading therefore becomes an estimate, the accuracy of which depends on the accuracy of the scale, the ability of the operator to read the scale, and in some cases the sensitivity of touch or feel on the part of the operator.

1.221 Scale Errors

If the scale against which a measurement is made is in error, then obviously that measurement will be in error. This can only be overcome by calibrating the instrument scale against known standards of length over its whole length.

In comparative measurements the effects of scale errors are reduced by using as short a length of scale as possible, by choosing a setting master whose size is as close to that of the gauge being checked as is conveniently possible.

1.222 Reading Errors

How accurately can a scale be read? This depends upon the thickness of the rulings, the spacing of the scale divisions and the thickness of the datum or pointer used to give the reading.

As a guide, a reading of a pointer or datum line against a scale division can be taken as having an accuracy of $\pm 10\%$ of the scale division. On the other hand the estimate of the position of a pointer between the rulings will be less accurate and should be taken as $\pm 20\%$ of the scale division. Thus a reading of -3 units taken off a scale whose divisions represent 0·001 mm would represent a comparative measurement of $-0·003$ mm to an accuracy determination of $\pm 0·0001$ mm.

If, however, the reading had been $-3\cdot4$ scale units, then it would represent $-0\cdot0034$ mm $\pm0\cdot0002$ mm.

It must be realized that when a measurement is made with a comparator this type of error occurs twice, first when setting the instrument to a master gauge and again when the reading is taken on the workpiece.

1.223 Measuring Errors

The different types of error discussed above are cumulative, and in some cases a further amount must be added to allow for sensitivity of touch or feel. This will depend upon the type of instrument being used, and in general the effect is eliminated with comparators.

Consider now the problem of measuring the error in a plain plug gauge of nominal diameter 25 mm. The measurement is to be carried out using a comparator having a magnification of $5000 \times$ which is set to a gauge block of nominal length 25 mm having a known error of $-0\cdot0001$ mm to an accuracy of determination $\pm0\cdot0002$ mm. The comparator reading on the gauge block is 0 scale divisions and on the plug gauge $-1\cdot2$ scale divisions.

In this case the effect of elastic deformation can be ignored as the two parts are of similar material under similar pressure, although the conditions of contact are slightly different. The problem can be set out in a tabular manner as follows.

Error Element	Amount or Reading	Accuracy of Determination
Gauge block	$-0\cdot0001$ mm	$\pm0\cdot0002$ mm
Comparator setting	0	$\pm0\cdot0001$ mm
Comparator reading	$-0\cdot0012$ mm	$\pm0\cdot0002$ mm
Totals	$-0\cdot0013$ mm	$\pm0\cdot0005$ mm

Thus the gauge size is found to be 24·9987 mm but the accuracy of determination shows that it can be anywhere between the values of 24·9992 mm and 24·9982 mm. In fact the accuracy of determination is likely to be better than $\pm0\cdot0005$ mm (see page 11).

1.3 COMPOUND ERRORS

Many cases occur in which the measurement finally computed is a function of a number of individual measurements a, b, c, etc., all of which have individual accuracies of determination δ_a, δ_b, δ_c, etc.; then the accuracy of determination of M, which we can denote dM, could be found by substituting in the expression for M the maximum and minimum values of a, b, c, etc., and thus finding the maximum and minimum values for M.

This would obviously be a laborious process, and the problem is better solved using partial differentiation, whence it can be shown that

$$dM = \frac{\partial M}{\partial a} \cdot \delta a + \frac{\partial M}{\partial b} \cdot \delta b + \frac{\partial M}{\partial c} \cdot \delta c, \text{ etc.}$$

where $\frac{\partial M}{\partial a}$ is the differential coefficient of M with respect to a, all other variables being considered as constants in this term

$\frac{\partial M}{\partial b}$ is the partial differential of M with respect to b, etc.

Consider the problem on page 110, where it is shown that for a particular measurement,

$$D = L + \frac{W^2}{8L}$$

Let $L = 400$ mm and $\delta L = \pm 0{\cdot}025$ mm

$W = 50$ mm and $\delta W \pm 1{\cdot}00$ mm

$$dD = \left[\frac{\partial D}{\partial L} \cdot \delta L + \frac{\partial D}{\partial W} \cdot \delta W \right]$$

$$= \pm \left[\left(1 - \frac{W^2}{8L^2}\right) \delta L + \left(\frac{2W}{8L}\right) \delta W \right]$$

$$= \pm \left[\left(1 - \frac{50^2}{8 \times 400^2}\right) 0{\cdot}025 + \left(\frac{2 \times 50}{8 \times 400}\right) 1{\cdot}00 \right] \text{ mm}$$

$$= \pm \left[\left(1 - \frac{2}{512}\right) 0{\cdot}025 + \frac{1}{32} \right] \text{ mm}$$

$$dD = \pm [0{\cdot}024 + 0{\cdot}031] \text{ mm}$$

This solution has been deliberately left in two parts so that the significance of each part of the expression can be shown. Note that the error of $\pm 0{\cdot}025$ mm in L has a direct effect on the value of D, but the error of $\pm 1{\cdot}00$ mm in W only affects M by an amount of $0{\cdot}031$ mm, the point being that L must be determined to a far higher order of accuracy than W as dD contains 96% of δL but only 3% of δW.

1.4 THE EFFECT OF AVERAGING RESULTS

The accuracy of determination is the amount by which it is estimated that a measurement *could* deviate from its true value, but it is not necessarily true that it does so. By chance the measurement could possibly be exactly right or it could deviate from the true size by any fraction of the accuracy of determination. There

is no way of knowing other than using a method of measurement with a better accuracy of determination.

If we repeat the complete measurement a great many times, we should get a number of different values for the measured size x and we could obtain a frequency distribution by plotting a tally chart (see page 192) of these values. From this information we could calculate the standard deviation σ of the values (page 194). It is known that 99·28% of the observations will lie within $\pm 3\sigma$ of the mean of the observations, so that we can say that for practical purposes the estimated accuracy of determination is equal to $\pm 3\sigma$.

If now we take the observations and divide them into random sub-groups of n and for each sub-group calculate its mean size $\bar{x}$, we can produce a frequency distribution for the values of $\bar{x}$. A little thought will show that this distribution will be more closely grouped about the true mean size than that for the individual sizes. It can be shown that

$$\sigma_m = \frac{\sigma}{\sqrt{n}}$$

where σ_m = standard deviation of the means $\bar{x}$

  σ = standard deviation of the individual observations

  n = sub-group or sample size

It follows that the accuracy of determination of the mean size of a sample of n observations is

$$\pm 3\sigma_m \quad \text{or} \quad \pm \frac{3\sigma}{\sqrt{n}}$$

Statistical tables show that approximately 95% of all observations lie within $\pm 2\sigma$ of the mean of the observations, and approximately 65% lie within the limits of $\pm 1\sigma$ of the mean. Hence we can now state the confidence with which we give the accuracy of determination. Let the estimated accuracy of determination of a single observation be $\pm \delta$. As this represents $\pm 3\sigma$ we can say that we are confident this accuracy of determination will hold good for more than 99% of all such observations. More simply, we say that $\pm \delta$ represents the 99% confidence limits. Similarly, $\pm 2\sigma = \pm \frac{2}{3}\delta$ giving 95% confidence limits and $\pm 1\sigma = \pm \frac{1}{3}\delta$ giving 65% confidence limits.

If we apply this to the mean size of n observations, we see that

$$99\% \text{ confidence limits} = \pm \frac{\delta}{\sqrt{n}} \qquad 95\% \text{ confidence limits} = \pm \frac{\frac{2}{3}\delta}{\sqrt{n}}$$

$$65\% \text{ confidence limits} = \pm \frac{\frac{1}{3}\delta}{\sqrt{n}}$$

To do this with accuracy we should have to take a large number of observations and from them calculate the true value of the standard deviation, but by estimating the accuracy of determination we can not only give its value but also the approximate degree of confidence we can assign to that value.

1.5 GRAPHICAL METHODS

If an experiment is carried out to find the law relating two measured variables, x and y, it is usual to plot a graph of the readings and determine the law of the graph by plotting a mean line, i.e. we are averaging out the errors in the individual observations. Usually the observations are manipulated to give a straight line graph of the general form $y = ax + b$, where a is the slope or gradient of the line and b is the intercept on the y axis. If the line is drawn on the graph in a position estimated by eye, then this will still be liable to error. If at each point a rectangle is drawn representing the accuracy of determination of the individual observations, then two lines can be drawn through the extremes and two laws calculated, these giving the limits of the accuracy of determination of the derived law. This is obviously a tedious process, and a better method is known as the *method of least squares*.

1.51 Method of Least Squares

Consider the slope a of the graph. This may be expressed as the *average* increase in y for a given increment of x, and it can be expressed as

$$a = \frac{\Sigma(x - \bar{x})(y - \bar{y})}{\Sigma(x - \bar{x})^2}$$

In practice this expression for a necessitates a laborious calculation. A general expression which simplifies calculations by enabling the original data to be used without first calculating the means is

$$a = \frac{\Sigma xy - \dfrac{\Sigma x(\Sigma y)}{n}}{\Sigma x^2 - \dfrac{(\Sigma x)^2}{n}} \qquad \ldots (1)$$

Having found the best value for a, the best value for b can be found by substituting *average* values for x and y in the expression

$$\bar{y} = a\bar{x} + b \qquad \ldots (2)$$

where $\bar{y} = \dfrac{\Sigma y}{n}$ and $\bar{x} = \dfrac{\Sigma x}{n}$

The problem is best set out in tabular form, and from the above expressions we see that we shall need columns for the observed values x and y and further columns for x^2 and xy. The total values Σx, Σy, Σx^2 and Σxy will also be required.

Consider the experimental values for x and y below:

	1	2	3	4	5	6	7	8	9	10	11	12	13	Totals
x	1	2	3	4	5	6	7	8	9	10	11	12	13	91
y	17	18	24	31	33	37	33	36	41	44	57	57	54	481
x^2	1	4	9	16	25	36	49	64	81	100	121	144	169	819
xy	17	36	72	124	165	222	231	288	369	440	627	684	702	3966

From (1) above,

$$a = \frac{3966 - \dfrac{481 \times 91}{13}}{819 - \dfrac{91 \times 91}{13}} = \frac{3966 - 3367}{811 - 637}$$

$$\therefore \quad a = 3 \cdot 29$$

$$\bar{y} = \frac{481}{13} = 37 \qquad \bar{x} = \frac{91}{13} = 7$$

Substituting for $\bar{x}$ and $\bar{y}$ in expression (2),

$$37 = (3 \cdot 29 \times 7) + b$$
$$\therefore \quad b = 13 \cdot 97$$

The law of this particular straight line is then

$$y = 3 \cdot 29x + 13 \cdot 97$$

If now we substitute the values of x in the above expression, we obtain the theoretical values of y which we may denote Y. The errors in the observed values are then $(y - Y)$. This calculation requires two extra columns in the tabular calculation; $y = ax + b$ and $(y - Y)$.

It has been suggested* that this method should be used in the evaluation of straightness and flatness tests.

1.6 SUMMARY

The main points arising from this chapter may be summarized as follows:

(*a*) All measurements are subject to error.

(*b*) The possible deviation from the stated measurement should be estimated and given as an accuracy of determination.

(*c*) The accuracy of determination can be improved by repeating the measurement a number of times and stating the mean value.

* A. J. Scarr, *Proc. I. Mech. E.* Vol. 82, part 1, no. 23. 1967–8.

(*d*) Statistical methods of expressing the accuracy of determination, and the confidence with which it is stated, are available and should be understood and used.

(*e*) Statistical methods may also be applied to the analysis of experimental data which are normally expressed graphically, and in general these methods give a better fit of the experimental data to the laws derived therefrom.

In the section dealing with accuracy of determination and measuring errors (on page 6) the accuracy of determination calculated represents the worst possible case. In the example quoted there are three elements to the measurement:

(*a*) the gauge block

(*b*) the comparator setting

(*c*) the comparator reading

each one having an accuracy of determination $\pm \delta_1$, $\pm \delta_2$ and $\pm \delta_3$. The accuracy of determination calculated is the arithmetic sum of the individual accuracies $\pm (\delta_1 + \delta_2 + \delta_3)$.

In fact it must be realized that δ_1, δ_2 and δ_3 all have equal chances of being plus *or* minus, and the probability of any or all of them being at a maximum value is small. Statistically a better estimate of accuracy of determination is given by:

$$\text{accuracy of determination} = \pm \sqrt{(\delta_1{}^2 + \delta_2{}^2 + \delta_3{}^2)}$$
$$= \pm 0 \cdot 0001 \sqrt{(2^2 + 1^2 + 2^2)}$$
$$= \pm 0 \cdot 0003$$

Similarly, in the example on page 7, the best estimate of accuracy of determination, dM, is given by:

$$dM = \pm \sqrt{[(0 \cdot 024)^2 + (0 \cdot 031)^2]}$$
$$dM = \pm 0 \cdot 039 \text{ mm.}$$

CHAPTER 2

Light Waves as Standards of Length

2.1 THE EVOLUTION OF A LENGTH STANDARD

It is fundamental to the science of measurement, and hence the degree of control which it exerts on the development of technologies, that it should be based on an agreed, and if possible internationally agreed, system of standards. For many years the major industrial countries of the world used two systems, imperial and metric.

The disadvantages of such an arrangement are evident when one considers that a very large part of the world's population used metric units, but that important industrial countries, notably the United Kingdom and the U.S.A., used both imperial and metric units, the former being dominant in the industrial field. Virtually the only concession made by these countries was that scientific work was carried out in metric units.

The basis for a solution to this confusion was established in 1960 when the General Conference of Weights and Measures, an international body, recommended that SI units should be brought into use to replace existing metric units. SI is an abbreviation of Système International d'Unités (International System of Units) which has grown out of the MKS (metre, kilogramme, second) system and the MKSA (metre, kilogramme, second, ampere) system. The major industrial countries, including those at present using a metric system, have adopted the recommendation. Thus the United Kingdom is at present in the process of conversion from imperial to SI units. The process will take some years for its completion.

The standard of length, therefore, will be the metre, and for the purposes of this book will be the most important of the SI units considered.

2.11 The Metre Defined

As part of the evolution of a universal standard of length, the International Committee of Weights and Measures recommended in 1958 that the metre be defined as

$$1\ 650\ 763{\cdot}73 \times \lambda$$

where λ = the wavelength, in a vacuum, of the orange-red radiation of the isotope krypton 86

12

It is this definition which has been universally adopted by those countries using, or intending to use in the future, SI units.

Clearly, a universal standard must be one which is reproducible with such a degree of accuracy that for all industrial and scientific purposes it may be considered as absolute. By means of interferometry, the error of reproduction of the metre is of the order of 1 part in 100 million. Similarly, any subdivision of the metre may be produced, and reference to B.S. 4311: *Slip (or Block) Gauges and their Accessories* shows that the practical working standards of length used in industry are of such accuracy that the calibration and reference grades of these must be verified by interferometry, that is, in terms of the wavelength of light.

2.2 THE NATURE OF LIGHT

Interferometry is that branch of science which is concerned with the manner in which rays of light, produced from a common source, are recombined by a lens system, usually the eye. The difference in path lengths along which the rays travel before being recombined determines their phase relationship, and hence the sensation or otherwise, of light entering the eye.

For an understanding of the phenomena associated with interferometry, it is necessary to consider the nature of light.

Two theories have been advanced to explain the nature of light: the Emission Theory, and the Wave Theory. The former was advanced by Newton, and considered light as consisting of particles emitted by luminous bodies, the impact of the particles on the eye causing the sensation of light.

The wave theory, however, was advanced by Huygens, and considered light as a wave motion propagated in the ether.

It is this theory, and its subsequent development, which satisfactorily explains the phenomena associated with light, including that of interference.

If, then, light is considered as an electromagnetic wave of sinusoidal form, it may be represented as in Fig. 2.1.

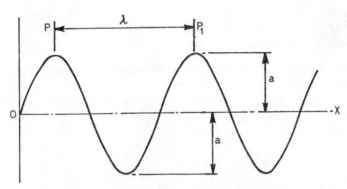

Fig. 2.1.   Light represented as a sine wave.

The direction of propagation is along the O–X axis, and the distance travelled by the wave during the time (T) occupied in one vibration, is the wavelength λ. That is, at points of equal disturbance such as P P_1, the distance between these points constitutes the wavelength. The maximum disturbance of the wave is the amplitude, a, and its intensity a^2. The velocity, v, of transmission $= \lambda/T$ and frequency $= 1/T$.

For any given monochromatic light source, these characteristics are virtually independent of ambient conditions such as temperature and pressure.

If we now consider, for example, the use of the green radiation from the spectrum of mercury 198 as a monochromatic light source for the absolute measurement of length, we have a wavelength of 0·5461 μm of such reproducibility and sharpness of definition, that length measurement of an accuracy of 1 part in 100 million may be made.

For many engineering purposes, white light, formed of the colours and therefore wavelengths of the visible spectrum, constitutes a valuable and convenient light source.

The following chart shows the approximate wavelengths of the colours forming the visible spectrum, and explains why such sources cannot be considered as being suitable as the basis of absolute length measurement. The individual colours cannot be sharply defined as being of definite wavelength, but for many practical measuring problems it is appropriate to consider the average wavelength, approximately 0·5 μm, of white light formed by the blending of the visible spectrum colours, as being sufficiently accurate to constitute a standard of length.

Colour	Range of Wavelengths (μin)	Range of Wavelengths (μm)
Violet	15·7–16·7	0·396–0·423
Blue	16·7–19·3	0·423–0·490
Green	19·3–22·6	0·490–0·575
Yellow	22·6–23·6	0·575–0·600
Orange	23·6–25·4	0·600–0·643
Red	25·4–27·5	0·643–0·698

That the primary colours have such ill-defined wavelengths is the principal reason for the intensive efforts made by physicists over many years to produce pure monochromatic light such as that from mercury 198 or krypton 86, having a precise, reproducible wavelength.

2.3 MONOCHROMATIC RAYS

A ray of monochromatic light may be considered as being composed of an infinite number of waves of equal wavelength, the value of which determines the colour of the light.

If we now consider the effects of combining two such rays, we may do so by considering only two waves, one from each ray.

They may combine in such a way that the resultant effect is that the maximum amount of illumination is obtained. This condition is shown in Fig. 2.2 in which

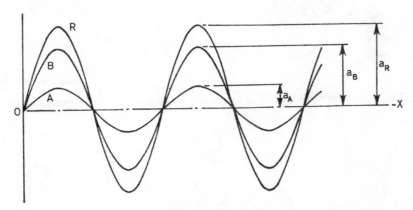

Fig. 2.2. Resultant amplitude a_R of two waves A and B of different amplitudes a_A and a_B, but in phase.

two rays, A and B, are in phase at their origin O, and clearly will remain so for any distance of propagation.

If A and B have the amplitudes shown, then the resultant wave R will have an amplitude $a_R = a_A + a_B$.

In other words, when A and B are of the same wavelength and in phase, they combine to increase the amplitude, and therefore the intensity, of the resultant R to a maximum.

If now we consider the effect of changing the phase relationship of A and B by the amount δ, as in Fig. 2.3, it can be shown that when $a_A = a_B$, $a_R = 2a \cos \delta/2$.

That is, the combination of A and B no longer produces maximum illumination.

Further, if we consider the effect of changing the phase relationship of A and B, so that they are displaced 180°, or $\lambda/2$, then a condition as in Fig. 2.4 occurs, in which R has an amplitude which is the algebraic sum of a_A and a_B and is reduced to a minimum. It is important to note that if a_A and a_B were equal in value, then a_R would be zero, as $\cos (180/2) = 0$. That is, complete interference between two waves of the same wavelength and amplitude produces darkness, and no sensation of light is registered on the retina of the eye.

It is the ability to split light from a single source into two component rays, to recombine them, and observe the way in which they recombine, that allows the

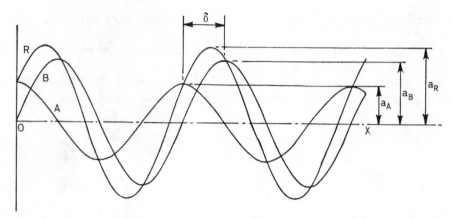

Fig. 2.3. Resultant amplitude a_R of two waves A and B out of phase by an amount δ.

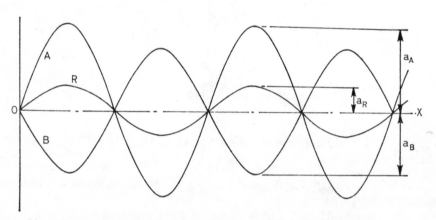

Fig. 2.4. Resultant amplitude a_R of two waves A and B out of phase by half a wavelength (180°).

wavelength of light to be used for linear measurement. It should be noted that it is the linear displacement, δ, between two waves which is being measured, and that the maximum interference between them occurs when $\delta = \lambda/2$.

Interference of two rays of light may be demonstrated in the following way. Referring to Fig. 2.5, A and B are effectively two point sources of monochromatic light having a common origin. S is a screen the plane of which is parallel to the line joining A and B. $O-O_1$ is perpendicular to the screen, and intersects the line AB at its mid-point.

The rays from A and B, since they have a common origin, and are therefore of the same wavelength, will be in phase.

If we now consider the effect at the point O_1 on the screen, of the converging of the two rays from the point sources A and B, it is clear that since $AO_1 = BO_1$

16

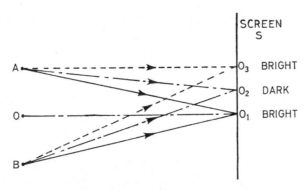

Fig. 2.5. Formation of alternate light and dark areas on a screen, due to waves from sources A and B travelling different path lengths.

the two rays will arrive at O_1 in phase, and will combine in a manner similar to that shown in Fig. 2.2 to give the maximum illumination at O_1.

If we now consider a point such as O_2 on the screen, the distance AO_2 is clearly less than the distance BO_2, and if $BO_2 - AO_2$ is equal to an odd number of half wavelengths, that is $(2n+1)\,(\lambda/2)$, where n is an integer, then the waves will be 180° out of phase, complete interference will occur and there will be darkness at this point.

Again, if we consider a point such as O_3, then if $BO_3 - AO_3$ is equal to an even number of half wavelengths, that is $2n\,(\lambda/2)$, then the rays are again in phase, and no interference occurs at O_3, this point receiving maximum illumination.

This process could be repeated at points both above and below O, and would result in alternate dark and light areas being formed. The dark areas would occur wherever the path difference of A and B amounted to an odd number of half wavelengths, and the bright where their path difference amounted to an even number of half wavelengths.

It must be emphasized that the phenomena described would occur only if the sources A and B were images of a single source. This can be achieved with a Fresnel biprism by which light passing through a slit is divided into two identical and equally spaced images, the rays from which, emanating from the same source, will be in phase at the images. It is the difference in path lengths of the subsequent rays which causes interference.

Summarizing, it is clear that the following two conditions are necessary for light interference to occur.

(*a*) Light from a single source must be divided into two component rays.

(*b*) Before being combined at the eye, the components must travel paths whose lengths differ by an odd number of half wavelengths.

2.4 INTERFEROMETRY APPLIED TO FLATNESS TESTING

A manufacturing problem frequently encountered in precision engineering is the production of flat surfaces of relatively small area. Such surfaces are normally produced by grinding followed by successive lapping operations until a high degree of flatness combined with a high surface finish is achieved. Virtually the only satisfactory, and certainly the only convenient, method of testing the flatness of such surfaces is by the use of light interference, using an optical flat as a reference plane.

An optical flat is a disc of stress-free glass, or quartz. One or both faces of the disc are ground, lapped and polished to a degree of flatness not normally encountered on an engineering surface. For engineering purposes, therefore, the optical flat may be considered as a reference of flatness, and used as such for comparing engineering surfaces. Optical flats vary in size from 25 mm diameter to about 300 mm diameter, the thickness being about 50 mm for the largest. In all cases, they are relatively rigid and stress-free discs which, used and stored correctly, will retain their flatness and therefore effectiveness almost indefinitely.

If an optical flat is laid (not 'wrung') on to a nominally flat reflecting surface, it will not form an intimate contact, but will lie at some angle θ as in Fig. 2.6, in which θ is greatly exaggerated. A wedge-shaped air cushion may be formed between the surfaces.

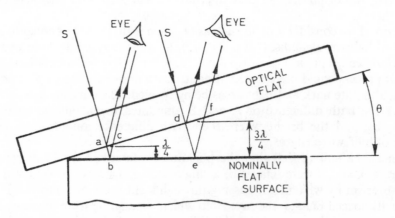

Fig. 2.6. Formation of interference fringes on a flat surface viewed under an optical flat in a parallel beam of monochromatic light.

If this arrangement is now placed in the path of a parallel beam of monochromatic light, we can consider S as the source of one wave of the incident beam. Ignoring any refractive effects due to the light passing through media, glass, and air, of differing densities, it is seen that the wave from S is partially reflected at '*a*', and partially transmitted across the air gap, to be reflected at '*b*', a point on the work surface. The two reflected components are collected and recombined by the eye, having travelled paths whose lengths differ by an amount abc.

If abc$=\lambda/2$ where $\lambda=$wavelength of source, then the conditions for complete interference have been satisfied, i.e. the ray from S has been split into two components, and recombined; also, the path lengths of the components differ by an odd number (one) of half wavelengths.

If the surface is flat, then at right-angles to the plane of the paper it will be parallel to the optical flat, and these conditions will be satisfied in a straight line across the surface. Thus a straight dark line, or interference fringe, will be seen.

Further along the surface, and due to angle θ, the ray leaving S will similarly split into two components whose path lengths differ by an amount def.

If def$=3\lambda/2$, the next odd number of half wavelengths, interference will again occur and a similar fringe will be seen.

At an intermediate point, the path difference will be an even number of half wavelengths, the two components will be in phase, and a light band will be seen at this point.

Fig. 2.7.   Interference fringes on a flat surface viewed under an optical flat in a parallel beam of monochromatic light.

Thus, a surface will be crossed by a pattern of alternate light and dark bands, which will be straight, for the case of a flat surface, as in Fig. 2.7. A deviation from straightness would be a measure of the error in flatness of the surface in a plane parallel to the apex of the angle θ.

Referring again to Fig. 2.6, it is seen that if θ is small (which it must be),

$$ab = bc = \frac{\lambda}{4}$$

and

$$de = ef = \frac{3\lambda}{4}$$

The change in separation between the optical flat and the surface is the difference between ab and de (or bc and ef).

$$\therefore de - ab = \frac{3\lambda}{4} - \frac{\lambda}{4} = \frac{\lambda}{2}$$

Thus the change in separation between the surface and the optical flat is equal to one half wavelength between similar points on similar adjacent fringes.

Note that if θ is increased the fringes are brought closer together, and if θ is reduced, i.e. the surfaces become more nearly parallel, the fringe spacing increases. The possible practical variation in θ is very small, since if the surfaces are too closely together ('wrung' together), no air gap exists, and no fringes are observable, and if θ is too large the fringes are so closely spaced as to be indistinguishable, and an observable pattern is not maintained.

In practice, it is unlikely that contact between the optical flat and the work surface will occur at one point only as in Fig. 2.6. It is probable that contact would be made at a number of points, or along one or a number of lines. Fig. 2.8 shows the pattern which would be observed if the work surface were spherically convex. Contact is made at the central high point, resulting in a fringe pattern of concentric circles. If, now, each adjacent fringe represents a change in elevation of the work surface relative to the optical flat of $\lambda/2$, then

$$\frac{\lambda}{2} \times n = \text{Total change in elevation from point}$$
$$\text{of contact to the outermost fringe}$$

where λ = wavelength of light used

and n = number of adjacent fringes observed

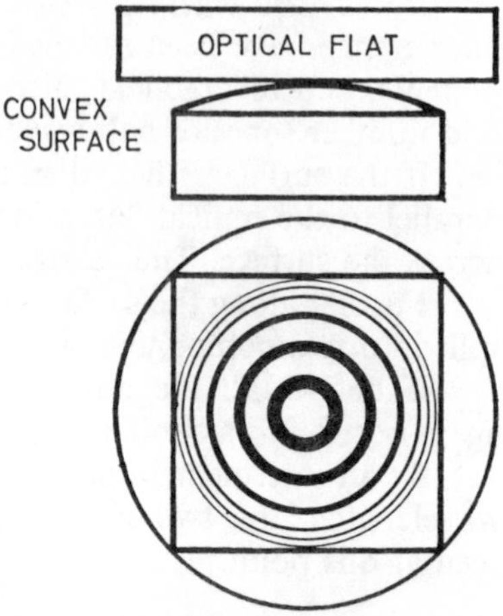

It is clear that the pattern observed when the surface is spherically convex will also be observed when the surface is spherically concave. To distinguish between these two conditions if, when the surface is spherically convex, one edge of the optical flat is lightly pressed, it will

Fig. 2.8. Fringe pattern produced on a convex surface. Note that as the angle between the surface and the optical flat increases the fringes become narrower and more closely spaced.

rock on a new high spot and the centre of the fringe pattern will move as shown in Fig. 2.9, and the outer fringes move closer together.

Also, when the surface is spherically concave, the flat rests on a line passing around the surface, and if the edge of the optical flat is lightly pressed, the edge line does not move as the pressure is varied. Alternatively, light pressure at the centre of the optical flat will cause it to deflect slightly and become more nearly parallel with the concave surface, thus reducing the number of fringes observed.

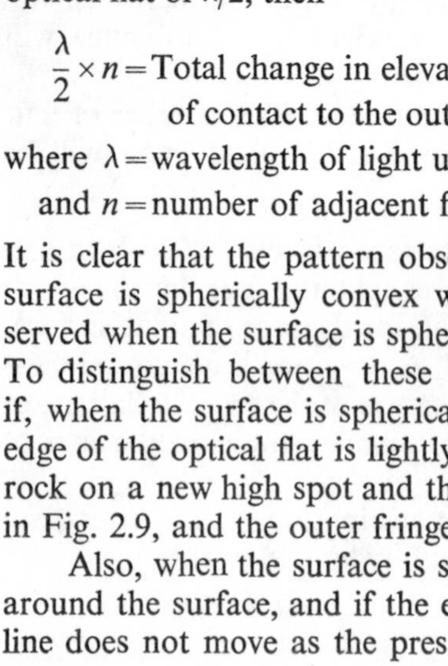

Commonly, optical flats are used in normal daylight, the spectrum of which has a wavelength of approximately 0·5 μm. Thus, each fringe interval corresponds to a change in elevation of the surface of 0·25 μm.

Suppose an optical flat to be laid on to a surface, and the resulting fringe pattern is that shown in Fig. 2.10.

Having first determined the point or line of contact of the optical flat, which is assumed to be at AA, it must be remembered that the contour of each fringe lies on points of equal height (in a negative direction) relative to the surface of the optical flat. Thus, the fringe pattern, in fact, represents a contour map of the surface

Fig. 2.9.   Method of testing to show that a surface is convex.

under test, the spacing of the fringes representing height intervals relative to the optical flat of $\lambda/2$. In Fig. 2.10, point C is the same distance from the optical flat as point B, but $\lambda/2$ farther (or nearer) than point D. Therefore the edge at C is $\lambda/2$ higher (or lower) than D. If the fringes curve towards the line of contact at A, the surface is convex, the opposite case also applying.

Practice in the use of optical flats is essential to a true understanding of the patterns produced, and at this point it may be appropriate to indicate the points to be observed in their use.

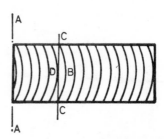

Fig. 2.10. Fringe pattern formed on a surface whose C.L. is half a wavelength higher than its edges.

(a) Handle optical flats carefully, and the minimum amount.

(b) Ensure that the work surface and the optical flat are perfectly clean, by careful wiping with a cloth of the 'Selvyt' type or with chamois leather.

(c) Never 'wring' an optical flat to a work surface. It should be laid on, so that the flat is not distorted by tending to adapt itself to the contour of the work surface, thus producing a false fringe pattern.

(d) Never 'wring' two optical flats together. Separation may be difficult and cause damage.

2.5 INTERFEROMETERS

Although optical flats can be used in either daylight, or, better, in a diffused source of near-monochromatic light, e.g. a light box consisting of a sodium discharge lamp behind a yellow filter, they suffer the following disadvantages for very precise work:

(a) It is difficult to control the 'lay' of the optical flat and thus orientate the fringes to the best advantage.

(b) The fringe pattern is not viewed from directly above and the resulting obliquity can cause distortion and errors in viewing.

These problems are overcome by using optical instruments known as interferometers, two of which will be discussed, one for measuring flatness and the other for determining the length of slip and block gauges by direct reference to the wavelength of light.

2.51 The N.P.L. Flatness Interferometer

This instrument, shown in diagrammatic form in Fig. 2.11, was designed by the National Physical Laboratory and is manufactured commercially by Coventry Gauge and Tool Co. Ltd., and Hilger and Watts Ltd.

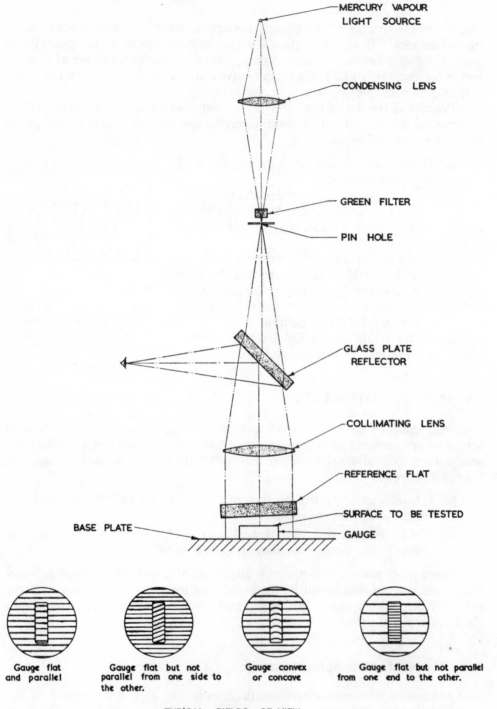

Fig. 2.11.   Optical arrangement of interferometer for testing flatness of surfaces.
(*Courtesy of the N.P.L. Crown Copyright*)

It consists essentially of a mercury-vapour lamp whose radiations are passed through a green filter, thus removing all other colours, and leaving a green mono-chromatic light whose wavelength is very close to 0·5 μm. This light is focused on to a pinhole, giving an intense point source of monochromatic light, which is in the focal plane of a collimating lens, and is thus projected as a parallel beam of light. This beam is directed on to the gauge to be tested via an optical flat so that interference fringes are formed across the face of the gauge, the fringes being viewed from directly above by means of a thick glass plate semi-reflector set at 45° to the optical axis.

It should be noted that the optical flat is mounted on an adjustable tripod, independent of the gauge base plate, so that its angle can be adjusted. Further, the gauge base plate is designed to be rotated so that the fringes can be orientated to the best advantage.

An advantage of this instrument is that it can also be used for testing the parallelism between gauge surfaces. Two methods are used:

(a) For gauges below 25 mm in length.

(b) For gauges greater than 25 mm in length.

When shorter gauges are used interference fringes are formed both on the gauge surface and the base plate. As the gauge is wrung on to the base plate its underside is parallel with its base plate. This means that if the gauge faces are parallel the fringes on the base plate should be equally spaced, and parallel with the fringes on the gauge surface.

If the gauge being tested is more than 25 mm in length the fringe pattern on the base plate is difficult to observe, but the base plate is rotary and its underside is lapped truly parallel with its working surface. Therefore if a non-parallel gauge is viewed the angle it makes with the optical flat will be as in Fig. 2.12 (a). If the table is turned through 180° the surface is now less parallel with the optical flat, Fig. 2.12 (b), and a greater number of fringes is observed.

Consider a gauge which exhibits 10 fringes along its length in one position and 18 fringes in the second position.

In Fig. 2.12 (a) the distance between the gauge and the optical flat has increased by a distance δ_1, over the length of the gauge, and in the second position [Fig. 2.12 (b)], by a distance δ_2.

It has been shown that the distance between the gauge and the optical flat changes by $\lambda/2$ between adjacent interference fringes.

$$\text{Therefore, } \delta_1 = 10 \times \frac{\lambda}{2}$$

$$\text{and } \delta_2 = 18 \times \frac{\lambda}{2}$$

The change in the angular relationship is $\delta_2 - \delta_1$.

$$\therefore \delta_2 - \delta_1 = 8 \times \frac{\lambda}{2}$$

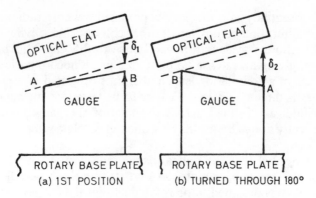

Fig. 2.12. Testing parallelism on gauges over 25 mm in length using flatness interferometer.

But due to the rotation through 180° there is a doubling effect. Therefore the error in parallelism $=\dfrac{\delta_2-\delta_1}{2}$.

$$\therefore \quad \frac{\delta_2-\delta_1}{2}=\frac{8}{2}\times\frac{\lambda}{2}=4\frac{\lambda}{2}$$

If the wavelength used is 0·5 μm, then

$$\frac{\delta_2-\delta_1}{2}=\frac{4\times0\cdot5}{2}=1\cdot0\ \mu m$$

Thus the gauge has an error in parallelism of 1·0 micro-metres over its length.

2.52   The Pitter–N.P.L. Gauge Interferometer

The mechanical subdivision of end standards of length tends to be laborious, especially when the smaller sizes are considered, and it is therefore liable to error. In view of this, and the requirements of B.S. 4311, it is no longer used for gauges less than 18 in or 50 cm in length, and has been superseded by interferometric methods of calibration.

A well-known interferometer is the Pitter–N.P.L. Direct Measurement Interferometer, which is based on a National Physical Laboratory design. Fig. 2.13 shows a diagrammatic arrangement of the instrument, and the field of view in the eyepiece.

In operation, light from the cadmium lamp is condensed by lens J on to the pin-hole plate G, thus providing an intense point source, in the focal plane of the collimating lens F. The resulting parallel beam passes to the constant-deviation prism E, where its colours, and therefore wavelengths, are split up and any one selected as required, by varying the angle of the reflecting faces of the prism relative to the reference plane B of the platen. The beam is turned through 90° and

24

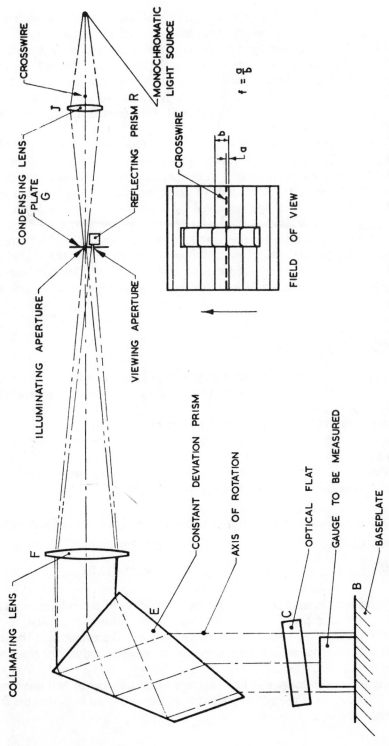

Fig. 2.13. Diagram showing the optical system for the 0–4 in N.P.L. interferometer for measuring slip gauges. (*Courtesy of the N.P.L. Crown Copyright*)

directed to the proof-plane C, the lower surface of which is coated with a semi-reflecting film of aluminium, which transmits and reflects equal proportions of the incident light. Part of the light is reflected upwards, and part is transmitted through the proof-plane to the upper surface of the slip gauge, and to the reference plane of platen B. Light reflected from each of these surfaces passes back through the optical system, but its axis is deviated slightly, due to the inclination of the proof-plane, so that it is interrupted by prism R to be turned through 90° into the eyepiece and the observer's eye.

The fringe pattern observed may be as shown. One set of fringes is due to the reflecting surface of the platen, and superimposed on this are the fringes due to the upper surface of the slip gauge wrung to the platen. Generally, the sets of fringes will be displaced as shown, the amount of displacement varying for each colour, and therefore wavelength, of light resolved by the constant-deviation prism. The displacement observed, a, is expressed as a fraction of the fringe spacing, b, i.e. $f = a/b$. It is sufficient to estimate this fraction, but to assist in this the cross-wire may be moved across the optical path, its image appearing in the eyepiece. An estimation of f is made for each of the four radiations from the cadmium lamp, red, green, blue, violet, resolved by suitably rotating the constant-deviation prism.

2.521 *Method of Measurement*

It is important to bear in mind that the physical conditions surrounding measurements of a nature such as this must be standardized and controlled. The standard conditions are as follows:

20°C temperature; 760 mm Hg barometric pressure; with water vapour at a pressure of 7 mm Hg and containing 0·03% by volume of carbon dioxide. If the conditions of measurement vary from these, correction factors must be applied.

Consider the measurement of a 3·0 mm slip gauge, using three wavelengths of the cadmium radiation. The wavelengths used will be as follows:

Red	0·643 850 37 μm	or, in 1 mm, there are	3106·311 80 half wavelengths		
Green	0·508 584 83 μm	,, ,, ,,	,, ,,	3932·480 64	,,	,,
Violet	0·467 817 43 μm	,, ,, ,,	,, ,,	4275·172 05	,,	,,

Fig. 2.14 (*a*) shows the arrangement, diagrammatically, of the reference plane or platen, slip gauge, and proof-plane, while Fig. 2.14 (*b*) shows the fringe relationship, $a/b = 0·65$ for the red radiation, as viewed through the eyepiece of the instrument. For each of the wavelengths employed, in succession, a different fraction a/b will be observed.

The height G of the gauge will be equal to a whole number of half wavelengths, n, plus the fraction a/b of the half wavelength of the radiation in which the fringes are observed.

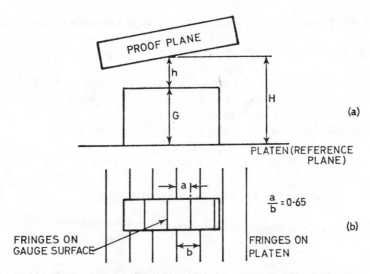

Fig. 2.14. Fringe displacement between gauge and base plate in N.P.L.-type gauge interferometer.

i.e. $G = H - h = n \cdot \dfrac{\lambda}{2} + \dfrac{a}{b} \cdot \dfrac{\lambda}{2}$

We thus have observed fractions a_1/b_1, a_2/b_2, a_3/b_3, for each of the radiations, which may be expressed as $f_1 : f_2 ; f_3$ respectively.

A series of expressions may now be obtained for the height of the gauge above the platen.

$$G = \frac{\lambda_1}{2}(n_1 + f_1); \quad G = \frac{\lambda_2}{2}(n_2 + f_2); \quad G = \frac{\lambda_3}{2}(n_3 + f_3)$$

But the values of f_1, f_2, etc., are the observed fractions a/b and not necessarily the values which would be obtained by calculation using the nominal height of the gauge and the values of λ for the three radiations.

We therefore have, for the nominal size:

$$G_N = \frac{\lambda_1}{2}(N_1 + F_1)$$

$$G_N = \frac{\lambda_2}{2}(N_2 + F_2)$$

$$G_N = \frac{\lambda_3}{2}(N_3 + F_3)$$

where G_N = nominal gauge height

N = number of whole half wavelengths in G_N

F = fractional displacement of fringes for any given radiation and due to height G_N in which N and F are found by dividing the half wavelength $\lambda/2$ into the nominal height of the gauge G_N

By combining the two sets of equations, we obtain the general expression:

$$G - G_N = \frac{\lambda}{2}[(n - N) + (f - F)]$$

Assume that the observed fractions, f, are $f_1 = 0.23$; $f_2 = 0.33$, and $f_3 = 0.71$, and that the calculated values of F are $F_1 = 0.94$; $F_2 = 0.44$; $F_3 = 0.52$ for the three radiations red, green, and violet respectively.

Inserting this information in the above equations we have:

$$G - G_N = \frac{0.643}{2}[(n_1 - N_1) + (0.23 - 0.94)] \ \mu m$$

$$= \frac{0.508}{2}[(n_2 - N_2) + (0.33 - 0.44)] \ \mu m$$

$$= \frac{0.467}{2}[(n_3 - N_3) + (0.71 - 0.52)] \ \mu m$$

$$\therefore \ G - G_N = \frac{0.643}{2}[(n_1 - N_1) + (0.29)] \ \mu m \ \ Note: (1 - 0.71) = 0.29$$

$$= \frac{0.508}{2}[(n_2 - N_2) + (0.89)] \ \mu m \ \ Note: (1 - 0.11) = 0.89$$

$$= \frac{0.467}{2}[(n_3 - N_3) + (0.19)] \ in$$

The values $(n_1 - N_1)$, $(n_2 - N_2)$ and $(n_3 - N_3)$ are unknown but it is known that they are (*a*) whole numbers and (*b*) relatively small numbers.

They can be found by trial and error, and it is found that if

$$(n_1 - N_1) = 2$$
$$(n_2 - N_2) = 2$$
$$(n_3 - N_3) = 3$$

a closely similar result for all three equations is found. However, this is laborious and a better method is to set the information out in tabular form as follows:

1	2	3	4	5	6	7
Wave-length (λ) μm	*Observed Fractions* (f)	*No. of* $\lambda/2$ *in* 3.0 mm	*Calculated Fractions* (F)	*Col. 2−Col. 4*	*Scales Coincide*	*Mean error in Gauge Length*
$R = 0.643$	0.23	9318.9354	0.94	0.29	2.29	
$G = 0.508$	0.33	11 796.441 92	0.44	0.89	2.84	0.74 μm
$V = 0.467$	0.71	12 825.5165	0.52	0.19	3.13	

The values shown in columns 6 and 7 are obtained by the method of scale coincidence. This offers the simplest method of obtaining the whole numbers of half wavelengths and fractions of half wavelengths which are contained in the error in the gauge. A slide rule may therefore be used (Fig. 2.15) in which the wave-

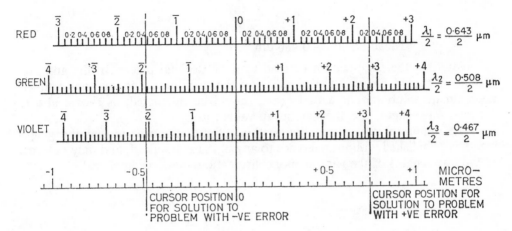

Fig. 2.15.   Layout of scales of half wavelength on slide rule for use with N.P.L. gauge interferometer.

lengths of red, green, and violet are set out to scale, from a common zero. The values obtained in column 5 of the table are found to have a close degree of coincidence at the values shown in column 6. The lower scale of the slide rule has graduations corresponding to micro-metres and a line passing through the values 2·29, 2·89 and 3·19 on the wavelength scales cuts the linear scale at 0·74 μm, the error in the height of the gauge.

It is found in practice that the values of f cannot be read to an accuracy greater than about 0·05 of a fringe spacing. This accounts for the small discrepancies in the fractional parts of the values shown in column 6, compared with the fractions shown in column 5. Since the linear error resulting from an observational error of this order will always be much less than 0·02 μm, it can be tolerated.

It should be noted that the half wavelength units on the negative side of the scales are $\bar{n}$ not $-n$, i.e. the whole numbers are negative but the fractions are still positive. Thus considering the cursor line shown to the left of zero, the displacements are respectively:

$$\text{Red} \quad \bar{2}\cdot48 = (-2 + 0\cdot48) = -1\cdot52 \text{ half wavelengths}$$
$$\text{Green} \quad \bar{2}\cdot09 = (-2 + 0\cdot09) = -1\cdot91 \quad ,, \qquad ,,$$
$$\text{Violet} \quad \bar{3}\cdot92 = (-3 + 0\cdot92) = -2\cdot08 \quad ,, \qquad ,,$$

Thus the gauge error is equal to those values multiplied by the respective half wavelengths.

$$\therefore \text{Error} = \begin{cases} -1 \cdot 52 \times \dfrac{0 \cdot 643}{2} \mu m = -0 \cdot 488 \ \mu m \\[2mm] -1 \cdot 91 \times \dfrac{0 \cdot 508}{2} \mu m = -0 \cdot 485 \ \mu m \\[2mm] -2 \cdot 08 \times \dfrac{0 \cdot 467}{2} \mu m = -0 \cdot 486 \ \mu m \end{cases}$$

Mean error $= -0 \cdot 486 \ \mu m$

Note that the scale cannot be read to three decimal places. In the authors' opinion it should be used for establishing the position of coincidence and an approximate solution. The actual errors should be calculated, as shown above, for each wavelength, and then the mean value found.

The important fact about this method of measurement is that the gauge length is established without reference to any physical standard, and only in terms of the wavelengths of the various monochromatic radiations employed.

CHAPTER 3

Linear Measurement

3.1 GENERAL

WE have been concerned in previous chapters with the methods by which engineering standards of length are established, and why such standards are necessary. This field is outside the normal scope of engineering metrology, and is more properly within the province of the physicist.

Such work, however, is the necessary basis of the linear measurement carried out by the engineer, since almost invariably this takes the form of comparing the size of a workpiece or other part with the known size of an end gauge, i.e. comparative measurement.

To carry out such measurements successfully, that is, to the order of accuracy required, often calls for ingenuity in the use of relatively simple equipment. It always requires the application of certain simple principles, together with patience, a systematic approach to a measurement problem, and the use of techniques only acquired by practice. These points are perhaps best illustrated by the apparently simple task of establishing the size of a plain gap gauge. There can be no substitute for experience in carrying out such a measurement.

3.2 SLIP AND BLOCK GAUGES

These are the engineer's practical, working, standards of length, and are remarkable for several reasons.

The following table is extracted from B.S. 4311, and specifies the accuracy of dimensions, flatness and parallelism of slip gauges.

The table does not, however, reveal the complete characteristics of slip gauges. In addition, they have a very high degree of dimensional stability, and possess the ability to be 'wrung' together. Dimensional stability is brought about by the careful selection of the steel from which they are made, and the stabilizing heat-treatment process the gauges undergo after hardening. The property of 'wringing' is due to the flatness and finish of the defining surfaces being such that when two gauges are brought into intimate contact, not only is the air film between them virtually removed, but molecular attraction occurs between the surfaces. The latter may be shown by the fact that when two 'wrung' gauges are left for a period of time, the difficulty of separation may be considerable. In fact it is advisable not to leave gauges in the wrung condition for longer than necessary, otherwise damage to the surfaces may result.

It will be noted that slip gauges are available in five grades of accuracy. For all workshop purposes, and for a wide range of inspection work, grade II is of sufficient accuracy. Grade I is appropriate to the checking and setting of gauges, and the standardizing of comparators. Grade 0 is suitable for the highest quality of work involving the verification of size of workshop and inspection limit gauges, while grade 00 should only be referred to on those occasions when the accuracy of the other gauges is in dispute, or requires verification. They exist in the background, as it were, of an inspection system as the ultimate practical standard. This extract from B.S. 4311: 1968 *Metric Gauge Blocks* is reproduced by permission of the British Standards Institution, 2 Park St., London W1A 2BS, from whom copies of the complete standard may be obtained.

Table 1. Tolerances on Flatness, Parallelism and Length of Gauge Blocks

As altered Feb., 1969

Size of gauge block		Calibration Grades and Grade 00				Grade 0			Grade I			Grade II		
				L										
	Up to and inclu-			Calibra-tion Grade	Grade 00									
Over	ding	*F*	*P*			*F*	*P*	*L*	*F*	*P*	*L*	*F*	*P*	*L*
mm	mm													
—	20	5	5	±25	±5	10	10	±10	15	20	+20 −15	25	35	+50 −25
20	60	5	8	±25	±8	10	10	±15	15	20	+30 −20	25	35	+80 −50
60	80	5	10	±50	±12	10	15	±20	15	25	+50 −25	25	35	+120 −75
80	100	5	10	±50	±15	10	15	±25	15	25	+60 −30	25	35	+140 −100

Tolerances. Unit = 0·01 micrometer (0·000 01 mm)

L = gauge length, *F* = flatness, *P* = parallelism

The practical value of all grades of slip gauges is much enhanced by a table of errors derived by the calibration of each individual gauge. Usually, the calibration is carried out by the gauge makers, or by an authority such as the National Physical Laboratory, using interferometric methods as dealt with in Chapter 2.

It is clear from such calibration that, although the smallest nominal difference between the sizes of two gauges is 0·0025 mm, it may in fact be possible to arrange differences appreciably smaller than this, by noting the amount and direction of error of individual gauges.

Where the quality of work demands it, the calibrated values of slip gauges should always be used. Normally 'protection slips' of 1·0 mm, and made of tungsten carbide, are provided with each set. These should be used on the outside of any combination to prevent undue wear of the actual gauges.

3.3 LENGTH BARS

The full specification of length bars is given in B.S. 1790, and provides for four grades of accuracy; workshop, inspection, calibration and reference.

Their purpose is similar to that of slip gauges, but for larger sizes of work. They are therefore less commonly found in the average engineering works, or inspection department. The end faces of the individual bars are lapped to a 'wringing' finish but, due to their weight and general proportions, this property cannot be relied upon to satisfactorily hold them in combination.

The bars, except those of 25 mm in length, are therefore threaded internally at each end, to allow their end faces to be brought into secure contact by a free-fitting screwed stud. Normally, any combination of the bars will have the 25 mm bars at each end to provide flat defining surfaces for the total length.

Length bars are best used in the vertical plane but when used horizontally they should be supported at two points equidistant from each end, the distance between the support points being $0.577 \times$ length of bars in the combination. The end faces are then brought mutually parallel, the axis of the combined bars being horizontal outside the support points. It will be noted that this is the same condition of support as shown in Fig. 1.2 (*a*).

It cannot be too strongly emphasized that the accuracy of slip gauges and length bars (the latter where large dimension work is involved), should form the foundation upon which an engineering company builds the accuracy of its products, and the interchangeability of component parts.

This requires the regular verification of the accuracy of gauges. It is not necessary for a company to possess a length interferometer with which to verify a reference grade set of gauges. This work is undertaken by the National Physical Laboratory as a service to industry. For example, at intervals of perhaps two years a grade 00 set of slips may be sent for verification, and a Certificate of Examination obtained. This will show the length errors in the individual gauges, and whether they lie within permitted tolerances of flatness and parallelism. With these gauges as reference standards, and used in conjunction with a high magnification comparator of suitable repeatability of reading, secondary sets of gauges may be calibrated for use in the inspection department. From these, workshop standards may be verified by a similar method.

There is thus established a direct relationship between the accuracy of work-pieces and the ultimate standards of length. It is the only rational basis for inter-changeable manufacture, and is set out in chart form in Fig. 3.1.

It is, of course, necessary to carry out the verification of gauges under suitable physical conditions. The reference gauges should be permanently housed in a standards room, located in a quiet area, and not subject to vibration from outside sources. Temperature control to plus or minus 1°C is desirable, and may be associated with humidity control. Dust control, through the filtering of air entering the room, is a further precaution against the deterioration of equipment.

All comparators should be permanently mounted on rigid supports, and in a level position. Given these conditions, and a technique of careful measurement, it is possible to determine the accuracy of gauges to a fraction of a micro-metre. The appropriate technique is dealt with in association with the design and operation of the comparators suitable for this class of measurement.

Primary standard of length (metre)
(Established by interferometry)
|
Secondary standards
(Verified by interferometry)
|
Grade 00 or Calibration grade slip gauges
(Verified by interferometry)
|
Grades 0 or I slip gauges
(Verified by high magn. comparator)
|
Grade II slip gauges
(Verified by high magn. comparator)
|
Workpiece
(Verified by suitable gauging method)

Fig. 3.1.

3.4 DESIGN AND OPERATION OF LINEAR MEASUREMENT INSTRUMENTS

Measuring instruments and machines incorporate, in their important features, principles which are based on kinematics.

Kinematics may be defined as a study of the relative motions of parts, in isolation from the forces which produce the motions.

It will be seen that, in general, the motions which must be considered in the design of measuring machines are those of straight line and rotary motion. At times, it may also be necessary to eliminate all such motions between parts, again through the application of kinematic principles.

Design experience over many years has shown that kinematic principles must be closely followed, in order that machines and instruments should possess the following characteristics.

(*a*) A high degree of sensitivity.

(*b*) A high degree of accuracy.

(*c*) Freedom from variance.

(*d*) Minimum inertia in the moving parts of the indicating mechanism.

3.41 The Principle of Alignment

In addition to these points, which may be provided for by kinematics, a further important principle must be observed. This is the Principle of Alignment, which may be stated as: *The line of measurement, and the line of the dimension being measured, should be coincident.* This principle is so fundamental to good design that it is rarely departed from to any serious extent.

A simple example, and one of common experience, is found in the design of an external micrometer. In Fig. 3.2 it is clear that the principle is completely satisfied. This applies equally to an internal micrometer.

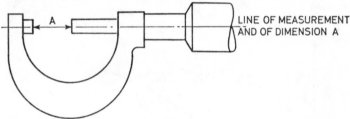

Fig. 3.2.   A hand micrometer conforming to the principle of alignment.

If, however, we consider a vernier calliper, it is also clear, Fig. 3.3, that the principle is not observed. It may also be shown, that as the displacement Y increases, the greater will be the probability of a discrepancy arising between dimension A and the reading B.

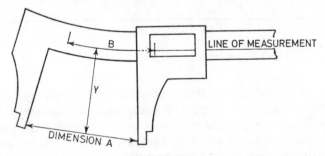

Fig. 3.3.   A vernier calliper does not conform to the principle of alignment.

It should be remembered that the curvature of the vernier beam may be brought about by the measuring pressure between the jaw faces. When one considers the extent to which this may vary from person to person (in the absence of any device to standardize it) it is seen that this accounts for the difficulty in establishing the dimension A with the same degree of confidence as with a micrometer.

Other examples dealing with the application of this principle will be examined later in the chapter.

We refer now to the other important functional characteristics which measuring machines and instruments should possess.

3.42 Sensitivity

This may be defined as the rate of displacement of the indicating device of an instrument, with respect to the measured quantity.

If we now consider sensitivity over the full range of instrument readings, and a graph is plotted of indicated readings with respect to measured quantities as in Fig. 3.4, the sensitivity at any value of $y = dy/dx$ where dx and dy are increments of x and y, taken over two consecutive readings. If continuous readings of y are taken over the full instrument scale, the sensitivity is the slope of the curve at any value of y.

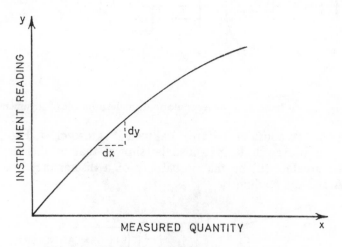

Fig. 3.4. Sensitivity of a measuring instrument.

It will be noted that this treatment of instrument sensitivity enlarges on the frequently accepted idea that it may be completely defined and determined by reference to the least change in the measured quantity which will cause an observable change in the instrument reading. Such a definition has no regard for the fact that sensitivity may not be uniform over the total displacement of the indicating device, such as a pointer.

3.43 Accuracy

This may be defined as the amount of correction which must be made to the instrument readings in respect of the values of the quantities being measured.

In practice, careful calibration of instrument scales will overcome this difficulty at the time of instrument assembly and testing but, in use, deterioration of the operating mechanism may set in, which will require recalibration of the instrument, to restore its accuracy. An analogy to this is the calibration of slip gauges to determine their actual size in relation to their nominal size.

3.44 Variance

This may be defined as the range of variation in instrument reading which may be obtained from repeated measurements of a given quantity.

Variance is inherent in many types of instruments, the extent to which it is present depending upon such factors as the quality of manufacture or the suitability of the operating principle for the type of measurement to be undertaken.

A simple and common example of variance is the operating characteristics of a dial test indicator.

A test to determine variance in such an instrument takes the form of a series of observations of indicated values in relation to the known values of slip gauges. The dial test indicator is rigidly mounted above a suitable reference surface, such as a lapped plate. A series of slip gauges is then selected, the series increasing in size in increments of say 1 mm and extending to embrace the full range of plunger movement. If then each slip gauge in turn is passed carefully across the reference surface and under the measuring plunger, a series of corresponding scale readings will be obtained, and recorded. Let us assume that these readings are taken with the plunger moving vertically upwards. The readings are then repeated, *but in the reverse direction*, and again recorded. If the readings are now

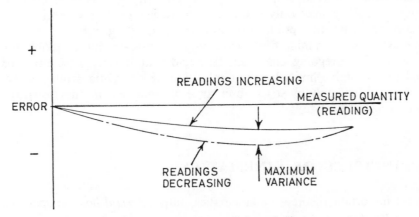

Fig. 3.5. Variance in a dial test indicator.

plotted as in Fig. 3.5, it may be found that the plotted points lie on two curves which form a loop, the boundaries of which represent the maximum variation of readings for a given value within the series of values of the slip gauges used. A pair of such curves forms a hysteresis loop, and is a characteristic of the instrument under test.

Fig. 3.5 shows that, in general, dial test indicators are not suitable for the measurement of height or length differences of a magnitude approaching the full range of indication. They are suitable for the measurement of small displacements, and in one direction only.

3.45   Inertia of Moving Parts

All instruments which depend wholly or in part on a linkage or other mechanical system, or on the displacement of a fluid, for their operation, are subject to the disadvantage of inertia. Only those instruments dependent solely on the application of optical principles are entirely free of inertia effects.

Inertia produces a condition referred to as passivity, or sluggishness. It may be determined for any given instrument by noting the smallest change in the measured quantity which produces any change in the instrument reading.

If we consider a diaphragm type of instrument in which the elastic movement of the diaphragm under fluid pressure is translated as, say a pointer movement relative to a scale, the instrument will be free of passivity only if the diaphragm is perfectly elastic under all variations of the fluid pressure to be measured.

The same characteristic applies to an instrument depending on a mechanical movement for the operation of a pointer. In this case, the imperfect elasticity of springs may be the source.

From the foregoing brief treatment of the sources of instrument errors, it will be appreciated that in the testing of instruments it may be difficult to separate a particular defect from others. For example, passivity is closely associated with sensitivity: passivity may only show itself as a change in the sensitivity of an instrument at a particular point in its scale reading.

It must be borne in mind that measuring machines, in addition to instruments, also possess the foregoing characteristics, and that in each case they may be reduced to acceptable limits by the application of kinematic principles. In fact, it may be said that only by strict observance of them can the functional requirements of a design be satisfied.

3.5   PRINCIPLES OF KINEMATICS

The most important theorem of kinematics states: *A rigid body in space has six degrees of freedom.*

To explain this, consider a body in space, Fig. 3.6. It has freedom of translatory motion along any one of the mutually perpendicular axes X, Y, and Z, and may also rotate about any one of these, thus making a total of six degrees of freedom. To completely fix the body in space therefore, six constraints must be applied; one constraint for each degree of freedom.

Commonly, in measuring instruments and machines, it is necessary to allow one degree of freedom of a member, which requires five constraints, or to completely constrain a member, it thus constituting a fixture.

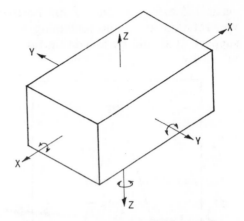

Fig. 3.6. Six degrees of freedom. If a body is free to move in space, it may have one of, or a combination of any number of the motions shown.

3.51 Complete Constraint

As an example of the latter case, consider the mounting of a surface plate on a stand. The mounting must be carried out in such a manner that the following conditions are satisfied:

(a) The plate is completely constrained without the application of external forces.

(b) The plate must be free to expand and contract with changes of ambient temperature.

(c) The plate must be adequately supported to prevent its distortion.

(d) It must not be possible to accidentally dislodge the plate from its set position.

A further optional requirement may be that the mounting should make provision for the plate to be set in a true horizontal plane.

All of these conditions may be satisfied in the following way:

The underside of the plate is fitted with ball feet, spaced so that the centre of gravity of the plate is contained within the triangle formed by lines joining the centres of the balls (Fig. 3.7).

The upper surface of the stand is then fitted with three steel blocks: one containing a conical hole; another a vee groove; the third remaining as a flat surface. The arrangement of these is as shown in Fig. 3.8.

If now the ball feet are rested on to the hole, the vee groove and the plane, and we consider the plate as the body to be constrained, the conical hole will prevent all lateral movement along the X, Y, and Z axes. Rotation about the vertical Z axis, and one of the horizontal X or Y axes, is constrained by the vee, rotation about the remaining horizontal axis being restrained by the flat support. It should be noted that no clamping forces are applied, other than that of gravity.

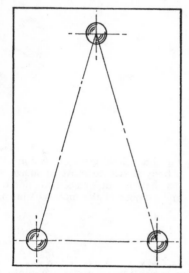

Fig. 3.7.   Underside of surface plate.

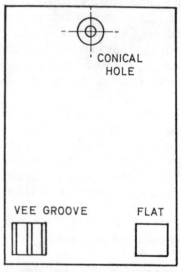

Fig. 3.8.   Plan view of kinematic mounting for surface plate.

Thus, the plate is completely constrained, and at the same time the other three conditions specified for its mounting are satisfied. It will further be noted, that the location of the plate is precise. As an additional refinement, if the three ball feet are carried on adjusting screws, their relative heights may be set so that the plate is brought to a true horizontal plane.

3.52   One Degree of Freedom

If we now consider the requirements for one degree of freedom, e.g. linear motion along the Y axis, if we substitute a second vee groove for the conical hole, we have a condition as in Fig. 3.9, the body being free to move only along the axis shown, provided its centre of gravity is contained within the triangle formed by lines connecting the centres of the ball feet.

A simple extension of this arrangement results in a slideway, by making the vee grooves continuous, as in Fig. 3.10.

For a slideway to be properly supported, and to reduce friction, a series of rolling balls are substituted for the ball feet, the distribution of which along the slide is ensured by pegs at suitable intervals. This type of slideway, or variants of it, is commonly found in measuring machines. The design of these will be dealt with under the applications of the particular machines. It is sufficient here to emphasize that only by the application of kinematic principles can the design of an instrument or machine be such that its accuracy in operation does not rest entirely on its accuracy of manufacture, although it must be realized that for the moving member of a vee–flat ball slide to have true linear motion the following conditions must be realized during manufacture:

(*a*) The vee must be straight in the horizontal and vertical planes.

(*b*) The flat must be straight in the horizontal plane.

(*c*) The vee and the flat must lie in parallel horizontal planes.

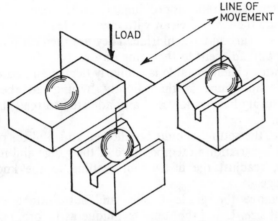

Fig. 3.9.   Kinematic constraints for linear motion.

Not only do kinematic principles allow simplicity of manufacture, but they provide for adjustment at instrument assembly and testing stage, so that completely satisfactory operating characteristics may be achieved.

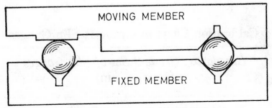

Fig. 3.10.   Vee–flat ball slide.

3.6 COMPARISON OF GAUGE BLOCKS

Mention has been made (section 3.3) of the technique suitable for the comparative measurement of slip gauges. It is of such fundamental importance that it is worth setting out in some detail. Although stated here as a procedure, it is clearly also a case where practical experience counts for much. The student, in particular, will profit considerably from carrying out an experiment in the calibration of slip gauges, involving the use of a high magnification, sensitive comparator.

Example. The calibration of a selection of grade II slip gauges, by comparison with grade 0 slip gauges.

Apparatus. A selection of grade 0 slip gauges. Grade II slip gauges to be calibrated. High magnification comparator ($\times 10\,000$) directly calibrated in units of $0 \cdot 10 \ \mu$m.

Method. It is assumed that the comparator and the gauges are housed in, or have been in, a standards room or metrology laboratory for sufficient time, so that equality of temperature at approximately 20°C exists between them.

The gauges are then wiped clean with chamois leather, after degreasing with a suitable spirit, say carbon tetrachloride. They are then set out into two groups, at a convenient point adjacent to the comparator.

The comparator platen or table is similarly degreased and cleaned.

A grade 0 slip gauge is then lightly 'wrung' to the platen and the comparator measuring head is lowered and adjusted to read the known error of the reference slip.

At this stage, the setting should be allowed to stabilize, principally to take account of possible variation in temperature of the gauge and its surroundings.

If necessary, readjust the instrument reading to the known error in the reference slip gauge.

Carefully remove the grade 0 slip, and substitute a grade II slip for the same nominal size. Using the same technique as before, note the error from the setting zero, and record it.

A precaution at this stage is to recheck the instrument datum, by taking a second reading on the reference slip. This is a check on possible changes in the measuring conditions, principally that of temperature.

Each grade II slip gauge may be similarly treated, and the results tabulated as below:

Calibration Chart of Grade II Slip Gauges

Standards: Grade 0 Set of Slip Gauges (Unit = 1 μm)			
Nominal Size (mm)	*Known error of Grade 0 Gauge*	*Reading on Grade II Gauge*	*Error in Grade II Gauge*
1	+0·1	+0·4	+0·4
2	0	+0·8	+0·8
3	+0·2	+0·5	+0·5
4	−0·1	+0·6	+0·6
5	+0·1	−0·5	−0·5

The accuracy of determination by such a method should be of the order of $\pm 0.05 \ \mu\text{m}$, but obviously it will vary with, and be dependent upon, the accuracy of the comparator, and the observer.

3.7 DESIGN OF COMPARATORS

Dimensional comparators are the principal instruments used in linear measurement, and as such their design has received much attention. Many principles have been used to obtain suitable degrees of magnification of the indicating device relative to the change in the dimension being measured. The main principles used are as follows:

(*a*) Mechanical.

(*b*) Mechanical–optical.

(*c*) Pneumatic.

(*d*) Electrical.

(*e*) Fluid displacement.

3.71 High-magnification Gauge Comparators

Additionally, comparators of particularly high sensitivity and magnification, and suitable for use in standards rooms, rather than inspection departments, have been designed, and brought into wide use for the calibration of gauges. These are:

(*a*) The Brookes Level Comparator.

(*b*) The Eden–Rolt 'Millionth' Comparator.

The design of each originated at the National Physical Laboratory, and was the work of men to whom much is owed in the field of fine measurement.*

Each of these instruments is now produced commercially, but the simplicity and soundness of their design is such that no book dealing with metrology can overlook them.

3.711 *Brookes Level Comparator*

The principle on which this operates is that of the displacement of the bubble of a sensitive level-tube, and is applied in the manner shown diagrammatically in Fig. 3.11 (*a*) and (*b*).

The important design features are the sensitive level-tube, and its mounting on a kinematic slide to allow it freely to take on the correct slope as required by the height difference of the gauges A and B, and the rotatable platen. This is flat

---

* A. J. C. Brookes, E. M. Eden, and F. H. Rolt.

on its upper and lower faces, which are mutually parallel. The lower face rests upon and may rotate on the flat upper face of the base. The arrangement is such that the platen is accurately rotatable in the plane of its upper face.

In use, the two gauges to be compared are 'wrung' to the platen, and the level-tube is lowered on its steel column so that the two ball feet, spaced at a centre distance of 17·5 mm, are resting on the upper surfaces of A and B. If then we assume that A is the reference gauge, the amount and direction of the inclination of the level-tube at position A–B will depend upon the height of B in relation to that of A.

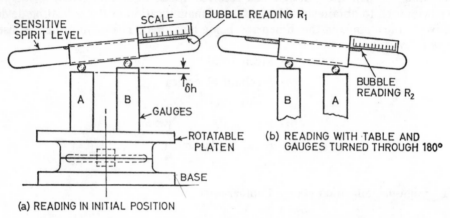

Fig. 3.11. Principle of comparative measurement by Brookes level comparator.

Also, since a level-tube gives readings in relation to a true horizon, account must also be taken of the plane of the platen in relation to this. This is done by rotating the platen through 180° to the position shown at Fig. 3.11 (*b*), having first raised the level-tube clear of the gauge. A second reading is taken at position B–A. The difference between the readings A–B and B–A must thus be twice the difference in height between the two gauges. We therefore have,

$$\text{Height difference } (\delta h) = \frac{R_1 - R_2}{2}$$

where R_1 and R_2 are the readings A–B and B–A respectively.

The instrument scale is calibrated directly to 0·2 μm, and may be read to 0·04 μm by estimation. It is thus possible, when one considers the 'doubling' effect of this method of comparison, to detect differences in height as small as 0·02 μm.

A feature of this instrument, and one which is quite unique, is that the gauges are compared together, and not in isolation from each other. The importance of this may be judged by the fact that having 'wrung' the gauges to the platen, they may then be left to equalize in temperature before readings are taken, a time of 1 min/mm of gauge length being recommended for this purpose.

The instrument is produced commercially in sizes to accommodate gauges up to 1 m in length. This again is a unique feature.

The operation of such an instrument is somewhat slow but, used under correct conditions in a standards room, is entirely suitable for the calibration of gauges, as dealt with previously in this chapter.

3.712 *Eden–Rolt 'Millionth' Comparator*

This instrument is again suitable for the calibration of gauges under measuring conditions appropriate to a standards room.

Its design is notable for the simplicity and economy with which it obtains a very high magnification, due to the combination of a mechanical movement supplemented by an optical system.

The mechanical magnifying system is shown diagrammatically in Fig. 3.12 (*a*).

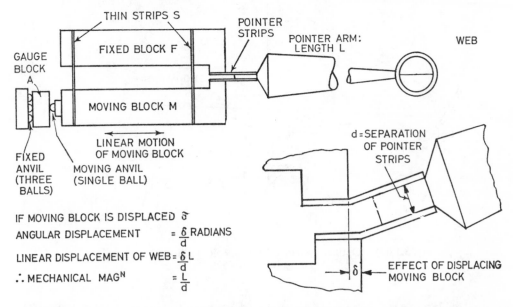

IF MOVING BLOCK IS DISPLACED δ

ANGULAR DISPLACEMENT $= \dfrac{\delta}{d}$ RADIANS

LINEAR DISPLACEMENT OF WEB $= \dfrac{\delta L}{d}$

$\therefore$ MECHANICAL MAG^N $= \dfrac{L}{d}$

Fig. 3.12(*a*). Mechanical system of Eden–Rolt 'millionth' comparator (magnification × 400).

The reference gauge, A, is placed between the anvil and the measuring plunger, and causes a small linear movement of block M relative to block F. The blocks are connected by thin steel strips, S, and to a pointer carrying at its other end a ring across which is stretched a spider web. The movement of block M causes a deflection of the pointer, which is about 200 mm in length.

The spider web at the end of the pointer is not viewed directly, but only as an image moving relative to a scale. The optical system producing this is shown diagrammatically Fig. 3.12 (*b*). It is a simple projection system, to give a magnification of × 50. If, then, the mechanical magnification = × 400, and optical magnification = × 50, total magnification = 400 × 50 = × 20 000.

The scale is calibrated to read directly to 0·2 μm and the scale divisions are about 5 mm apart. It is therefore possible to read by estimation to 0·02 μm.

It will be noted that this design produces a very high magnification in two relatively simple stages, the first of which depends upon kinematics for its sensitivity and simplicity of manufacture.

The contact faces of the anvil and plunger are also unusual. That carried on block M consists of a single ball. The anvil face, however, consists of a cluster of three balls enclosing a triangle. The gauge is therefore seated correctly on the anvil, and cannot take up an incorrect attitude relative to the line of movement of block M.

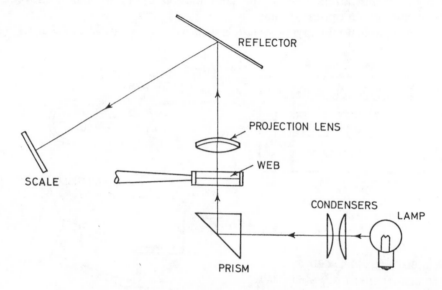

Fig. 3.12(*b*). Optical system of Eden–Rolt 'millionth' comparator (magnification × 50).

The instrument is considered to be an outstanding example of design, in which accuracy of manufacture plays little part in the accuracy of the readings. For example, it is not necessary for either the spacing of the steel strips carrying the pointer to be precise, or the length of the pointer. Similarly, the magnification of the optical system may be of a nominal value only.

Any deficiencies arising from the manufacture of either of these systems may be compensated for by the calibration of the scale of the instrument by reference to readings obtained on a series of slip gauges having known errors.

We will consider now the principles of operation of the more usual types of comparators, noting the ways in which kinematic principles are applied, and the instrument performance to be expected.

3.72   Mechanical Comparators

3.721   The Johansson 'Mikrokator'

Perhaps the simplest, yet most ingenious, movement used in this type of instrument is one due to H. Abramson, a Swedish engineer, and which is made by C. E. Johansson Ltd. It is shown diagrammatically in Fig. 3.13.

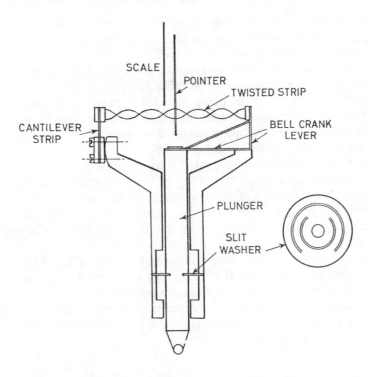

Fig. 3.13.   Movement of Johansson Mikrokator.

A thin metal strip carries at the centre of its length a very light glass tube pointer. From the centre, the strip is permanently twisted to form right- and left-hand helices. One end of the strip is fixed to the adjustable cantilever strip, the other being anchored to the spring strip elbow, one arm of which is carried on the measuring plunger.

As the measuring plunger moves, either upwards or downwards, the elbow acts as a bell crank lever and causes the twisted strip to change its length and thus further twist, or untwist. Hence the pointer at the centre of the twisted strip rotates an amount proportional to the change in length of the strip.

It can be shown that the ratio

$$= \frac{\mathrm{d}\theta}{\mathrm{d}l} = \text{amplification} = -\frac{9 \cdot 1 \, l}{W^2 n}$$

where l = length of twisted strip, measured along its neutral axis

W = width of twisted strip

n = number of turns

θ = twist of mid-point of strip with respect to the end (degrees)

The dimensions of the rectangular section of the twisted strip are always very small, and vary according to the amplification of the instrument. An average value for these dimensions is $0 \cdot 06$ mm $\times 0 \cdot 0025$ mm and the stresses in the strip, for a given tension, may be further reduced by small perforations along its length.

The purpose of the cantilever strip, other than as an anchorage, is to allow an adjustment to be made in amplification. Its effective length may be varied, and if for example it is increased, then for a given total movement of the plunger more of this movement will be accommodated by deflection of the cantilever, and less by extension of the twisted strip.

It should be noted that here again is an example in which design allows simplicity of manufacture; the final setting of the instrument amplification being made by a simple adjustment of the free length of the cantilever strip. It should also be noted that the cantilever mounting is adjustable, and by slackening one, and tightening the other mounting screw, the initial tension in the twisted strip may be adjusted.

The lower mounting of the plunger is in the form of a slit C washer as shown in Fig. 3.13 and this completes the movement of an instrument which has no mechanical pivots or slides at which wear can take place.

The instrument is surprisingly robust and is produced commercially in a range of magnifications up to $\times 5000$, although under controlled laboratory conditions much higher sensitivities have been achieved.

3.722 *The 'Sigma' Comparator*

Another mechanical comparator of ingenious yet simple design is that produced in a range of magnifications by the Sigma Instrument Company. The movement is shown in diagrammatic form in Fig. 3.14 (it must be emphasized that this is a diagram explaining the principles and that the actual movement is much more compact than this).

The plunger, mounted on a pair of slit diaphragms to give a frictionless linear movement, has mounted upon it a knife edge which bears upon the face of the moving member of a cross-strip hinge. This consists of the moving component and a fixed member, connected by thin flexible strips alternately at right angles to each other. It can be shown that if an external force is applied to the moving member it will pivot, as would a hinge, about the line of intersection of the strips.

Attached to the moving member is an arm which divides into a 'Y' form. If the effective length of this arm is L and the distance from the hinge pivot to the knife edge is x then the first stage of the magnification is L/x.

To the extremities of the 'Y' arm is attached a phosphor-bronze strip which is passed around a drum of radius r attached to the pointer spindle. If the pointer is of length R then the second stage of the magnification is R/r and the total magnification is $L/x \times R/r$.

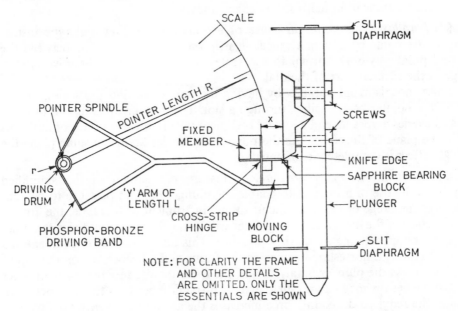

Fig. 3.14. Diagram of movement of Sigma mechanical comparator.

The magnification can be adjusted by slackening one and tightening the other screw attaching the knife edge to the plunger and thus adjusting distance x, while a range of instruments of differing magnifications can be produced by having drums of different radii r and suitable strips.

Apart from the movement the instrument has other interesting features as follows:

(a) *Safety*. It should be noted that the knife edge moves away from the moving member of the hinge and is followed by it. Therefore, if too robust a plunger movement is made the knife edge moves away from the hinge member and shock loads are not transmitted through the movement.

(b) *'Dead beat' readings*. The pointer is caused to come to rest, with little or no oscillation, by mounting on the pointer spindle a non-ferrous disc moving in the field of a permanent magnet. Rotational movement of the pointer and disc cause eddy currents to be set up in the disc, which have a turning effect in opposition to the direction of motion and proportional to the rotational velocity. Thus oscillations are damped out by a damping force which increases as required, i.e. with the amplitude and velocity of the oscillation.

(*c*) *Fine adjustment*. The normal height adjustment of the work stage, to zero the pointer, is too coarse for the high magnification of such an instrument, and is unsuitable for this purpose. To overcome this the dial of the Sigma instrument is mounted so that it rotates about the spindle axis. Thus, final setting of the pointer is achieved by moving the scale, rather than the work, and in this case the instrument magnification helps rather than hinders.

(*d*) *Parallax*. This common cause of erroneously reading dial type instruments is overcome by having a reflective strip on the scale. The line between the eye and pointer is only normal to the scale when the pointer obscures its own image in the silvered part of the dial.

Later models avoid parallax by having the pointer behind the scales, the tip only of the pointer being visible, through a slot. The pointer tip is turned through 90° and carries across its end a small 'Tee' piece which moves in the slot and thus lies in the plane of the scales. As the pointer and scales lie in the same plane the parallax effect is completely eliminated.

(*e*) *Constant measuring pressure*. A feature not shown in the simplified diagram is the use of a magnet to enable the plunger contact pressure to be constant over the range of the instrument. In any instrument of this type, the greater the deflection of the movement, the greater will be the displacement force required, and thus the greater the measuring pressure. This adverse effect is compensated for by mounting a horseshoe magnet on the frame and a keeper bar on the top of the plunger. As the plunger is raised, and the force required increases, the keeper bar approaches the magnet and the magnetic attraction between the two increases. Thus as the required deflecting force increases the assistance given by the magnet increases and the total force remains constant.

The Sigma comparator has been discussed at some length because it is felt that, apart from being an excellent measuring instrument, it provides an example of a first-class design technique, each problem being considered separately and a means of overcoming it devised, based on the use of relatively simple and well-known principles.

3.73 Mechanical–Optical Comparators

In these instruments, small displacements of the measuring plunger are amplified by, initially, a mechanical system consisting usually of pivoted levers, followed by a further amplification by a simple optical system involving the projection of an image. The mechanical system causes a plane reflector (mirror) to tilt about an axis. The image of an index is then projected to a position relative to a scale, and on to the inner face of a ground-glass screen.

Fig. 3.15 shows, in diagram form only, such an arrangement.

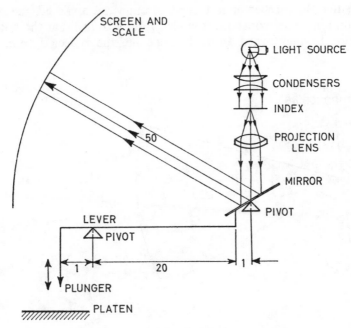

Fig. 3.15.   Principle of optical comparator.

In such a system:

Mechanical amplification $= 1 \times 20 \times 1 = 20$ units
Optical amplification $\quad = 50 \times 2 \quad = 100$ units
Total amplification $\quad\quad = 20 \times 100$ units
$\quad\quad\quad\quad\quad\quad\quad\quad\quad = 2000$ units

The factor of 2 contained in the optical amplification is brought about in the following manner.

Consider a parallel beam of light falling on to a plane reflector, and at angle θ relative to the normal to the reflector, Fig. 3.16 (*a*). The reflected beam will also form the angle θ relative to the normal.

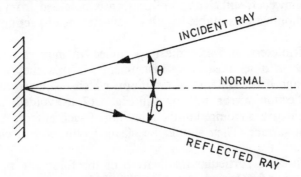

Fig. 3.16(*a*).   Reflection from a plane surface.

If now, the plane reflector is turned through the angle $\delta\theta$, the path of the incident beam remaining constant, it is seen [Fig. 3.16 (*b*)] that the normal to the plane reflector having turned through angle $\delta\theta$, the reflected beam is turned through $2\delta\theta$.

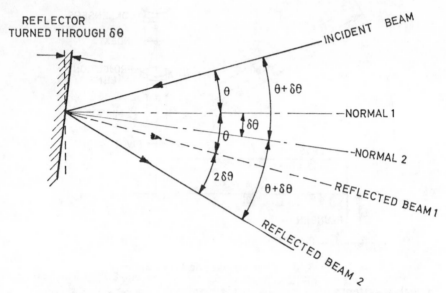

Fig. 3.16(*b*). Reflection from a plane surface turned through angle $\delta\theta$.

The angle between the incident beam and its new normal 2 equals $\theta + \delta\theta$. Likewise, the angle between normal 2 and reflected beam 2 is equal to $\theta + \delta\theta$. The total angle between the incident and reflected beams therefore becomes, $2(\theta + \delta\theta)$.

It is this 'built in' magnification of 2 obtained from a single reflection, which is of such value in optical comparators. Clearly also, if a double reflection is provided, as may be obtained from two opposed reflectors, one fixed, the other capable of tilting through $\delta\theta$, then the factor becomes 4.

Normally, however, sufficient amplification is obtained from a single mirror used in conjunction with a suitable length of reflected beam, acting as a weightless lever.

A point of interest, and of some importance, in mirrors used for this and similar purposes in other measuring instruments, is that they are of the front reflection type, and hence the loss of definition which results from the use of the normal back reflection mirror, as shown in Fig. 3.17, is avoided.

Considerable care is required in the handling of such mirrors to avoid damage to the reflecting surface. They should be cleaned only with a clean camel-hair brush.

Fig. 3.18 shows a particular application of the foregoing principles, in the Omtimeter produced by Optical Measuring Tools Ltd.

It is in effect an auto-collimator, in which the reflector is built in, and deflected by the measuring plunger. To keep the instrument compact in design the light path is turned through 90° by the prism, and of course the scale is graduated in linear units, usually of 0·001 mm.

Such instruments have very low inertia in the movements, as the only moving parts are the plunger and the reflector. The normal pointer in mechanical instruments is replaced by a beam of light which is of course weightless, and the optical lever system (Fig. 3.16) gives a doubling effect to the magnification.

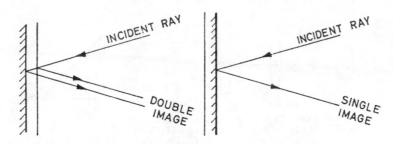

Fig. 3.17.   Reflection from back and front reflectors.

If the distance from the plunger centre-line to the mirror pivot is x, and the plunger moves a distance h, then the angular movement of the mirror

$$\delta\theta = \frac{h}{x}$$

If f is the focal length of the lens then the movement of the scale is $2f\delta\theta$ (see Chapter 4).

$$\therefore \text{ Scale movement} = 2f\frac{h}{x}$$

$$\text{The magnification of the instrument} = \frac{\text{Scale movement}}{\text{Plunger movement}}$$

$$= \frac{2fh}{x} \times \frac{1}{h}$$

$$= \frac{2f}{x}$$

It should be noted that this magnification is double that of a simple mechanical lever in which the magnification is the ratio of the lengths on either side of the fulcrum. In this case one lever arm length is the distance x; the other lever arm is the focal length f; and the factor 2 is due to the optical lever.

Further magnification is provided by the eyepiece so that:

$$\text{Overall magnification} = \frac{2f}{x} \times \text{Eyepiece magnification}$$

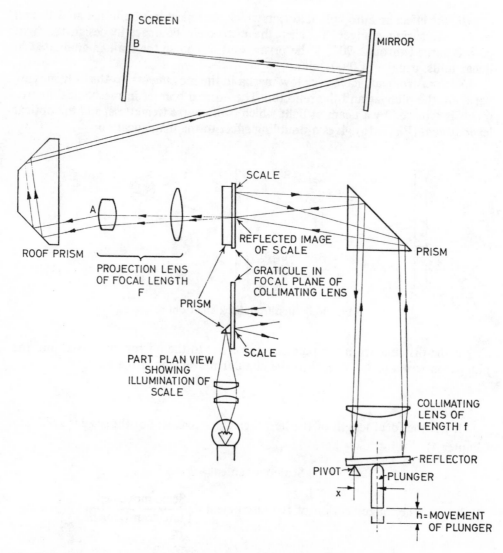

Fig. 3.18. Optical arrangement of O.M.T. optical comparator. This instrument projects the reflected image of the scale on to a screen.

$$\text{Magnification at graticule} = \frac{2f}{x}$$

$$\text{Projection magnification} = \frac{AB}{F}$$

$$\text{Overall magnification} = \frac{2f}{x} \times \frac{AB}{F}$$

(Courtesy of Optical Measuring Tools Ltd.)

3.74 Pneumatic Comparators

Industrially, pneumatic comparators, in which small variations are made in the dimension being measured with respect to a reference dimension and which are shown by a variation in either (*a*) air pressure, or (*b*) the velocity of air flow, are becoming of increasing importance.

The reasons for this are that very high amplifications are possible, no physical contact is made either with the setting gauge or the part being measured, and that internal dimensions may be readily measured, not only with respect to tolerance boundaries, but also geometric form. Further, the system lends itself to the inspection of a single, or a number of dimensions simultaneously, either during or immediately after the operating cycle of a machine tool.

3.741 Back-pressure Comparators

The air pressure variation system is based on the use of a two-orifice arrangement, as shown in Fig. 3.19.

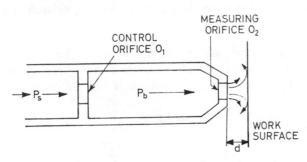

Fig. 3.19. Essentials of a back-pressure pneumatic gauging system.

Air is passed at controlled pressure into the measuring head, and provides the source pressure, P_s. It passes through the control orifice O_1 into the intermediate chamber. Orifice O_1 is of constant size, but the effective size of O_2 may be varied by the distance d. As d varies, pressure P_b also changes, and thus provides a measure of dimension d. Thus the indicating device is a pressure gauge or manometer recording the pressure P_b between the orifices.

By suitably matching the diameters of O_1 and O_2, and controlling P_s, the pressure at P_b may be made to vary linearly with the effective size of O_2, over a limited portion of the curve obtained by plotting the relationship of the ratios A_2/A_1 and P_b/P_s as shown in Fig. 3.20, where A_1 and A_2 are the areas of orifices O_1 and O_2 respectively.

For values of P_b/P_s between approximately 0·6 and 0·8, the curve is linear within 1%, and it is these values that are used in the design of such comparators for the relative diameters of orifices.

If we consider the linear portion of the curve, i.e. between the values of 0·6 and 0·8 for P_b/P_s its law may be written as:

$$\frac{P_b}{P_s} = a - \frac{bA_2}{A_1}$$

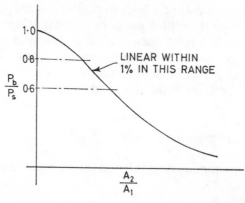

Fig. 3.20.   Characteristic curve of back-pressure pneumatic system.

As with any other comparator, the sensitivity is the ratio of the change of position of the indicator with respect to a corresponding change of position of the plunger. In this case the change of position of the indicator is dP_b and of the plunger dA_2, the pneumatic sensitivity being $\dfrac{dP_b}{dA_2}$.

$$\therefore P_b = P_s a - \frac{bA_2 P_s}{A_1}$$

$$\frac{dP_b}{dA_2} = \frac{-bP_s}{A_1}$$

Therefore, the pneumatic magnification is proportional to the input pressure, and inversely proportional to the area, or the square of the diameter, of the control orifice.

It is clear that an essential operating requirement is that pressure P_s is constant. It is thus necessary to have a simple pressure regulator controlling the pressure of the air from the normal supply line, and if necessary reducing it from about 55 N/cm² to 1 N/cm². Fig. 3.21 shows the circuit diagram of the instrument produced by Solex Air Gauges Ltd., the instrument being arranged for internal measurement.

The air from its normal source of supply, say the factory air line, is filtered, and passes through a flow valve. Its pressure is then reduced and maintained at a constant value by a dip tube into a water chamber, the pressure value being determined by the head of the water displaced, excess air escaping to atmosphere.

The air at reduced pressure then passes through the control orifice, and escapes from the measuring orifice. The back pressure in the circuit is indicated by the head of water displaced in the monometer tube. The tube is graduated linearly to show changes of pressure resulting from changes in dimension *d*, Fig. 3.19. Amplifications of up to 50 000 are obtainable with this system.

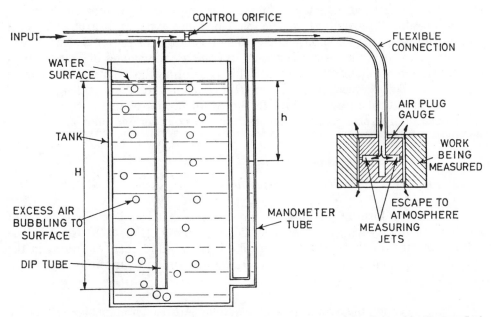

Fig. 3.21. Application of back-pressure air gauging system used by Solex Air Gauges Ltd.

Another back-pressure comparator is produced by Mercer Air Gauges Ltd., but this operates at the much higher pressure of 27·5 N/cm² gauge. The constant pressure input is produced from the line pressure by a diaphragm type regulator and passed to the control orifice and thence to the measuring orifice.

Interesting features are:

(*a*) *Magnification adjustment*. It has been shown that the magnification can be varied by varying the diameter of the control orifice. This is achieved by means of a taper-needle valve in the control orifice and enables a single scale to be used for all work by adjusting the magnification and zero settings.

(*b*) *Zero adjustment*. An air bleed, upstream of the measuring orifice and controlled by a taper-needle valve, provides a zero adjustment.

The pressure measuring device is a Bourdon tube type pressure gauge, the dial being graduated in linear units, i.e. 0·01 mm, 0·001 mm, or inch units.

As with all other comparators, initial setting is by means of reference gauges.

In this case, it is important that the reference gauges, and the part being measured, are of the same geometric form. For example, slip gauges are applicable as setting gauges for flat workpieces, while circular section work requires the use of cylindrical setting gauges. For work of the type shown in Fig. 3.21, a pair of reference ring gauges is necessary for setting purposes. If this precaution is not taken, the expansion characteristics of the air escaping from the measuring orifice, O_2, are changed and affect the accuracy of pressure readings on the manometer tube.

A possible disadvantage of the back-pressure type of instrument, is its relatively slow speed of response under some conditions of use. It is clear that as the volume of air in the system increases, its response to changes of pressure will, due to its compressibility, be reduced. This is a form of passivity referred to in section 3.45. This characteristic is of no great concern when the total length of the circuit is short, but in applications to dimensional control in the operation of machine tools for example, this length may be considerable, and give rise to passivity.

3.742 Flow-Velocity Pneumatic Comparators

The second form of pneumatic comparator is that based on the velocity of air flow. In this system, no control orifice such as O_1 (Fig. 3.19) is used. The velocity of air flow through the system is measured, and thus variations in the effective escapement area of the orifice O_2 are measured as variations in the flow rate of the air.

Fig. 3.22 shows the circuit for such a system. Air from a main supply, say the normal factory air line, is filtered and pressure regulated. It passes through an indicating device, in the form of a graduated glass tube, the bore of which is uniformly tapered and in which the indicator 'floats'. In fact the indicator takes up a position in the tapered tube such that the air velocity through the annulus created by the 'float' and the tube is constant. The air then escapes through the gauging orifice.

If, now, a master gauge of correct size and geometric form is used to provide a 'datum' rate of air flow through the system, and, as shown by the height of the indicator in the tube, any variation in the dimension of the part being gauged will produce a variation in the rate of flow through the system, by changing the effective cross-section of the gauging orifice. This is shown as a change in the height of the indicator on a linearly graduated scale.

This system may be used in all applications where the back-pressure system is applicable, and with the advantages of greater simplicity and speed of response, regardless of the length of the circuit.

Both the back pressure and velocity of air flow systems each have the advantage that the measuring orifice makes no physical contact with the surface of the part being measured. This is a considerable advantage where such contact may be detrimental to the surface finish of the part.

Additionally, each may be used to show the geometric form of a workpiece. This is a characteristic of very great importance, and will be increasingly so with the reduction of the size of dimensional tolerance zones. That is, the shape of a part must be contained within the diameters of two circles defining the tolerance boundaries. Where the differences between these is very small, perhaps a fraction of a micro-metre, the probability of a machine being incapable of producing geometric forms of the required accuracy increases.

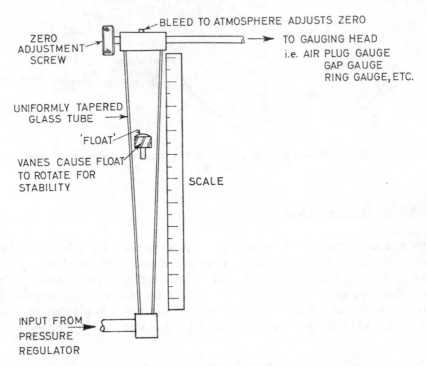

Fig. 3.22.   Line diagram of flow-velocity-type pneumatic circuit.

For example, the normal gauging unit for internal diameters consists of a free fitting plug gauge as in Fig. 3.21, having two orifices diametrically opposed. By rotating the head 180° in the bore being tested, the simple case of ovality may be detected.

By using a ring gauging head (Fig. 3.23) having three equally spaced orifices, the condition of a three-lobed shaft is detected, again by rotating the shaft through 120°. The same head will also give an average diameter of the shaft.

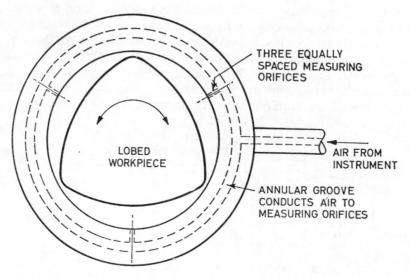

Fig. 3.23.   Pneumatic ring gauge arranged to measure diameter and reveal
lobing.

3.75   Electrical Comparators

In general these comparators depend upon the operation of a Wheatstone bridge
circuit in which, for a d.c. circuit, a change of balance of the electrical resistance
in each arm of the bridge is caused by the displacement of an armature, relative
to the arms, and under the action of the measuring plunger. The out-of-balance
effects in the circuit are measured by a micro-ammeter graduated to read in
linear units, and in terms of the displacement of the measuring plunger.

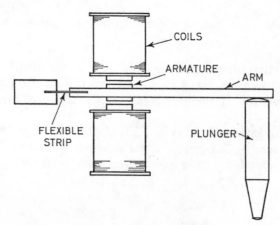

Fig. 3.24.   Measuring head for electrical comparator.

Fig. 3.24 shows a similar principle but for an a.c. circuit, in which the movement of the plunger displaces an armature, thus causing a variation in the inductance of a pair of coils forming one arm of an a.c. bridge. A possible arrangement for the measuring head of such an instrument is also shown, in which the plunger displaces an arm mounted between the coils on a thin flexible steel strip. The arm carries an armature, and the inductance in the coils is dependent upon the displacement of the armature relative to the coils. The balance of the circuit is initially set to zero, and the amount of unbalance caused by the movement of the measuring plunger, and hence by that of the armature, is amplified and shown on a scale graduated in linear units. Magnifications as high as $\times 30\,000$ are possible with this system.

3.76 Fluid Displacement Comparators

At the present time, comparators employing this principle, Fig. 3.25, have limited application.

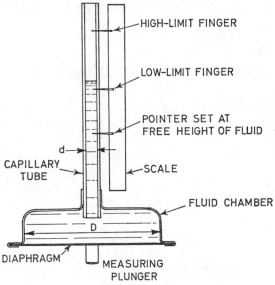

Fig. 3.25. Principle of Prestwich fluid gauge.

Magnification is approximately $\dfrac{D^2}{d^2}$.

A fine bore capillary tube is arranged so that its lower end is placed in a chamber of relatively large cross-sectional area containing a fluid of low viscosity. The bottom of the chamber takes the form of a diaphragm which may be deflected by pressure transferred to it by the measuring plunger. Diaphragm deflection causes a small quantity of fluid to be displaced from the chamber into the tube,

61

thus raising the level of the fluid in this to a point above its free height. A scale is arranged at the side of the tube, and limit pointers may be set relative to this to indicate the high and low limits of the dimension being measured, and as obtained by reference to a setting master.

The magnification of such a system is given by:

$$\text{Magnification} = \frac{\text{Cross-sectional area of chamber}}{\text{Cross-sectional area of tube}}$$

There are certain disadvantages in a system such as this:

(*a*) The fluid reacts to temperature changes in the same way as in a thermo-meter; hence the need for a pointer to show the free height in the tube. Any variation in this requires a corresponding resetting of the high- and low-limit pointers.

(*b*) There is a certain passivity in the instrument due to the characteristics of the diaphragm, and the viscosity of the fluid.

(*c*) The deflection of the diaphragm per unit measuring force is not constant. Hence, as the resistance offered by the diaphragm increases, the measuring force increases.

3.8 DESIGN AND OPERATION OF MEASURING MACHINES

The measuring instruments considered in section 3.4 are, with a few exceptions, limited to linear comparisons of about 250 mm. In fact several have capacities much less than this.

Much linear measurement work, however, involves dimensions considerably greater than this, and has given rise to a number of measuring machines of various designs. The simplest of these is based on the use of a micrometer screw as a means of revealing differences in length between a setting master and a workpiece or other gauge.

3.81 The Horizontal Length Comparator

The relatively simple machine of the type shown in Fig. 3.26 is a direct develop-ment of the bench micrometer, which is discussed in Chapter 8, dealing with screw thread measurement, and is suitable for length comparisons of up to 1 m.

It has all the main design features of the bench micrometer together with vee and flat guide-ways along the bed to allow suitable spacing of the measuring head, anvil, and work supports. Also the fiducial indicator is replaced by a dial gauge.

From 1.214 it has been seen that the correct support of bars, when held in a horizontal plane, is an essential feature of correct measurement, and that the two support points should be spaced $0.577 L$ apart, L being the length of the bar.

The supports shown in Fig. 3.26 are of vee form, and adjustable for height, spacing, and laterally across the bed.

In use, a length bar of appropriate grade, and as referred to in section 3.3, is placed on the vee supports which are then adjusted in height so that the axis of the bar and the axis of measurement are coincident, as shown by minimum readings of the dial indicator when one of the vee supports is adjusted vertically and across the bed, and a reading is obtained on the micrometer at some suitable position

Fig. 3.26. Length-bar measuring machine set to determine the length of a pin gauge.

of the pointer of the dial indicator. The gauge or workpiece to be measured is now substituted for the length bar and a second micrometer reading is taken at the same reading of the indicator. The difference between the two micrometer readings is clearly the difference between the lengths of the setting standard and the work, allowance being made for the known error in the length bar as obtained from the calibration chart. Since the micrometer reads direct to 2 μm, an accuracy to at least this order may be expected in determining the length of the workpiece, but it should be noted that whenever possible the length of the setting master should be close to the length of the workpiece to avoid undue movement of the micrometer screw and the consequent introduction of its pitch errors into the measurement. It is also necessary to allow a suitable period for temperature stabilization before taking any readings.

3.82 Universal Measuring Machines

Several machines of this type are available to cover a wide range of types of measurement. One such machine, Fig. 3.27, is made by Société Genevoise. The basis of the accuracy of measurement is an accurately divided scale viewed through a microscope, and used to determine the movements of the measuring anvils of the machine. It should be noted that the divided scale is positioned to be in line with the line of measurement as determined by the axis of the measuring anvils. Thus, the principle of alignment (section 3.41) is satisfied.

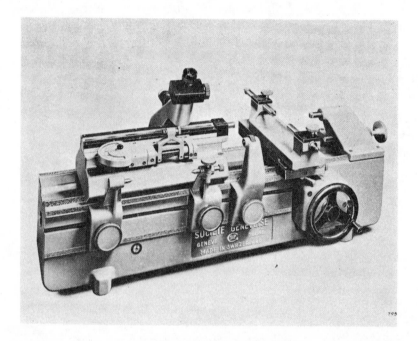

Fig. 3.27.   Measuring machine incorporating a divided scale as a reference.
(*Courtesy of the Société Genevoise*)

The machine is universal in the sense that lengths, and diameters, of both plain and threaded work, tapers, and the pitch of screw threads may be measured, and to a high order of accuracy.

Similar machines are built by Société Genevoise, employing two and three co-ordinate measuring systems, also based on the use of line standards. It is of interest that the scales are enclosed within the machines and are thus not liable to damage by physical contact and atmospheric attack, as are the working faces of end gauges.

3.83   The Photo-Electric Microscope

The reading accuracy of measuring machines based on the use of built-in line standards has been greatly enhanced by the development of the photo-electric microscope. The scale reading in such a machine is taken by causing a pair of hairlines in a microscope graticule to straddle evenly a unit line on the scale, and combining the scale and graticule readings. It must be appreciated that even straddling by the hairlines is a matter of subjective judgment and is thus liable to variation.

The photo-electric microscope overcomes this by scanning alternately, with a photo-transistor or cell, the light contained in the spaces between the straddling wires and the line. The electrical outputs from these alternate scans are caused to be of opposite sign and their algebraic sum is recorded on a meter. To make a reading therefore, the microscope is set by eye in the normal way and then adjusted until the meter reads zero, when the scale line is accurately positioned midway between the two hair lines. Thus greater repeatability of readings to a higher order of accuracy is obtained.

3.9   AUTOMATIC MACHINE CONTROL

It is felt that the topics of metrology and machine control are so closely associated that some mention of the latter must be made here, although in fact machine control is the subject of other volumes and papers, and space limits a detailed consideration of the subject.

Consider an operator performing a cylindrical grinding operation. To obtain a workpiece of the correct size he must carry out the following sequence of operations:

(*a*) Measure the work.

(*b*) Calculate the difference between the actual and the required sizes.

(*c*) Move the table cross-slide by half this amount.

(*d*) Take the cut set.

(*e*) Check by measurement the final size:

In an automatic sizing device this procedure is carried out continuously, a gauging head measuring the work continuously and transmitting its readings to a computor which compares the size at a given instant with the final size required. The computor output is then used to control the infeed of the grinding wheel.

A block diagram of this system is shown in Fig. 3.28. It is immediately apparent that the measuring unit of such a system is most important, and further that the measuring unit must have an output capable of being used as the input to the computor, i.e. it must read in terms of a physical quantity rather than a numerical value. For this type of work the back-pressure type of pneumatic gauging system is extremely useful and is incorporated into automatic sizing devices produced by Mercer Air Gauge Ltd., and others.

Electrical output is also suitable for machine control purposes, and the output from a line standard and photo-electric microscope is used by Société Genevoise for tape control of their jig borers.

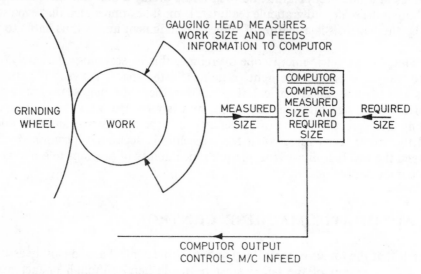

Fig. 3.28.   Essentials of an automatic sizing system.

Transmission diffraction gratings are used by Messrs. Ferranti Ltd. to provide a programmed control system on machine tools and other applications such as automatic flame cutting. The motion of Moire interference fringes across a pair of diffraction gratings is the measure of the movement of one grating relative to another. Thus, by counting the fringes, using a photo-transistor, the distance moved by a slide is compared with the distance the slide is required to move, and further motion is dependent upon this comparison.

In all cases, it should be noted, the whole system is dependent upon some method of linear measurement, either already existing, or being further developed to meet the demands of modern industry.

CHAPTER 4

Angular Measurement and Circular Division

4.1 GENERAL

ANGULAR measurement and circular division form a single and important part of metrology. Frequently, however, it is divided into two branches, angular measurement being concerned with the measurement of individual angles, while circular division, involving the continuous measurement of angles in the division of a circle, is considered as almost completely divorced from this system.

It is felt, however, that it is of such importance to appreciate these as two facets of a single problem, that this chapter will deal with them as such. Angular measurement will lead into circular division as a natural sequence.

4.2 ANGLES DERIVED FROM LENGTH STANDARDS—THE SINE BAR

It was noted in Chapter 1 that by suitable combination of linear measurements, angular measurements of a precise nature may be undertaken. The sine bar affords an excellent example of such combinations when used in conjunction with gauge blocks. In effect it consists of a bar carrying a suitable pair of rollers set at a known centre distance, the basic design and use being shown in Fig. 4.1.

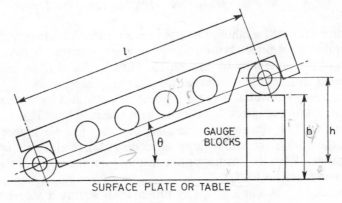

Fig. 4.1   Sine bar set up to angle *θ*.

If l is the linear distance between the axes of the rollers and h is the height of the gauge blocks, then sin $\theta = h/l$.

The design requirements of a sine bar are as follows, and unless these are carefully maintained the order of accuracy of angular measurement will fall:

(a) The rollers must be of equal diameter and true geometric cylinders.

(b) The distance between the roller axes must be precise and known, and these axes must be mutually parallel.

(c) The upper surface of the beam must be flat and parallel with the roller axes, and equidistant from each.

These requirements are met and maintained by care in the manufacture of all parts, which should be hardened and stabilized before grinding and lapping. Provided that all is in order, and with a 250 mm sine bar the distance between the roller axes should be accurate to within 1 μm, then it may be used in conjunction with gauge blocks to realize high orders of accuracy for setting individual angles.

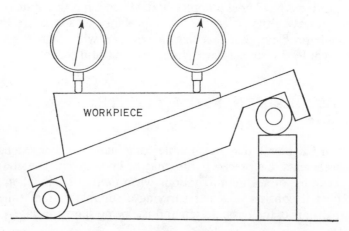

Fig. 4.2.   Sine bar used to check angle of wedge-shaped block.

In practice the sine bar should be used on a Grade A surface plate, and even so it is desirable to support both rollers on gauge blocks so that the minute irregularities of the plate may be eliminated. Thus in Fig. 4.1 the height h would be the difference in height between the two piles of gauge blocks.

To measure an individual angle care must be taken not to form a compound angle by having the workpiece misaligned with the sine bar. This is avoided by lightly holding the bar against an angle plate or cube. The workpiece is similarly held against the angle plate and a series of readings taken along its upper surface with a dial gauge. When the readings are constant then the angle of the workpiece can be obtained from sin $\theta = h/l$.

In determining a workpiece angle, rather than setting a sine bar to a predetermined angle, much time is often wasted in finding the correct value of h by trial and error. This time can be simply reduced by following a logical procedure as follows:

(*a*) Set up the sine bar and workpiece as in Fig. 4.2 so that the upper surface of the work is approximately parallel with the table surface.

(*b*) Take readings with a dial gauge at both ends and note their difference, noting which end of the work is low.

(*c*) Assuming that the end nearest the high end of the sine bar is low, then the gauge block height must be increased by an amount equal to the difference in the dial gauge readings multiplied by the proportion of sine bar length to work length.

For example, assuming that the end of a workpiece was 0·01 mm low, the sine bar being 250 mm long and the work 100 mm long, then the required increase in gauge-block height will be:

$$0·01 \times \frac{250}{100} = 0·025 \text{ mm}$$

This will not give an immediately correct setting from a first approximation, but it is much quicker than by a trial and error method.

Finally no sine bar should be used to set off angles greater than 45°, as beyond this angle the errors due to the centre distance of rollers, and gauge blocks, being in error, are much magnified. It is interesting to plot a graph of angular error against nominal angle if an error of 0·02 mm in all values of *h* is assumed. Large angles should be set off, where possible, by subtraction from 90°, i.e. set to the complement rather than the angle, and to a datum provided by an angle plate or cube known to be square to the table surface.

4.21 Sine Tables

These are a development of the sine bar principle and are set in a similar manner. It is clear that the sine bar is suitable only for relatively small work of light weight.

The sine table has a larger working surface and is much more robust than the sine bar and is suitable for larger, heavier work.

A further development is the compound sine table in which two sine tables, their axes of tilt set at right angles to each other, are mounted on a common base. The compound angle to be set is resolved into its individual angles in two planes at right angles to each other, and each table is set accordingly.

4.22 Sine Centre

For the testing of conical work, centred at each end, the sine centres shown in Fig. 4.3 are extremely useful since the alignment accuracy of the centres ensures that the correct line of measurement is made along the workpiece.

Metrology for Engineers

The principle of setting is the same as in the sine bar, although a hazard to be avoided is of the work and centres not being co-axial. To overcome this the work should be rotated on the centres until the maximum dial gauge reading is at the top. The angle is calculated from the gauge blocks in this condition, and then the work turned through 180° and the process repeated.

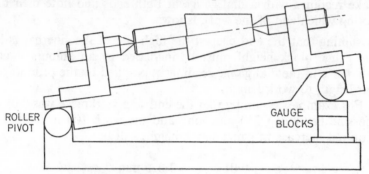

ROLLER
PIVOT

GAUGE
BLOCKS

Fig. 4.3.   Angle of a taper plug gauge ready for checking on sine centres.

The mean of the two angles determined will be the semi-angle of the work-piece, although it must be pointed out that any work which runs out to a measurable extent would probably be considered as sub-standard in quality and be rejected on this account.

4.3 MEASUREMENT OF TAPER GAUGES

Taper limit gauges are not normally used to measure the taper, or angle, in a hole or shaft, but rather to find whether the diameter of a particular cross-section of the work is within the allowed limits. The diameter checked is usually at one end of the work: therefore the gauge is made with a step, the diameter of the cone at the top and bottom of the step being the limiting sizes of the work.

The measurement of a taper plug or ring gauge may therefore be resolved into two stages:

(*a*) Determination of the angle of taper.
(*b*) Determination of specified diameters.

4.31   Measurement of Taper Plug Gauges

Usually the angle of taper of a taper plug gauge can be determined using the sine centres (section 4.22) but if these are not available the angle may be determined by making diametral measurements at known distances along the gauge.

The measurements are normally made over calibrated precision rollers of equal diameter, the separation of the individual measurements being controlled by gauge blocks. This arrangement is shown in Fig. 4.4.

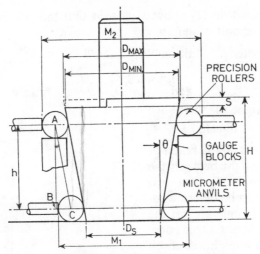

Fig. 4.4.   Measurement of taper plug gauge.

In triangle ABC it is seen that h is the height of the gauge blocks and $BC = (M_2 - M_1)/2$, in which M_1 and M_2 are the measurements over the rollers.

$$\tan \theta = \frac{BC}{AB} = \frac{M_2 - M_1}{2h} \quad \dots (1)$$

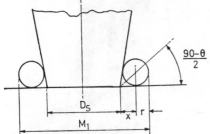

Fig. 4.5.   Measurement of the diameter of the small end of a taper plug gauge.

Thus the semi-angle of taper can be determined by direct measurement.

To find the maximum and minimum diameters at the top and bottom of the step it is first necessary to determine D_s, the diameter at the small end. Fig. 4.5 represents the small end of the gauge during the measurement of M_1.

It is seen that

$$M_1 = D_s + 2r + 2x$$

and from Fig. 4.5

$$\tan \frac{90 - \theta}{2} = \frac{r}{x}$$

$$\therefore x = r \times \text{cotangent} \; \frac{(90 - \theta)}{2}$$

$$\therefore M_1 = D_s + 2r \left(1 + \cot \frac{90 - \theta}{2} \right)$$

$$\text{or } M_1 = D_s + d \left(1 + \cot \frac{90 - \theta}{2} \right) \text{ where } d = \text{roller dia.}$$

$$\therefore D_s = M_1 - d \left(1 + \cot \frac{90 - \theta}{2} \right)$$

71

Referring to equation (1) above it is seen that tan θ is in fact equal to the increase in radius per unit of length.

$\therefore$ Increase in diameter per unit length $= 2$ tan θ

$$\therefore D_{max} = D_s + 2H \tan \theta$$
$$\text{and } D_{min} = D_s + 2(H - S) \tan \theta$$

where H is the height of the gauge and S is the height of the step

If a measurement M_3 is taken at some intermediate height between the positions for M_1 and M_2 then two further values of θ can be determined. If all these values of angle are not the same then the taper is not a true cone, i.e. its sides are not straight. A check on roundness can also be made by carrying out the measurements M_1, M_2, and M_3, at different positions around the gauge.

4.32 The Taper Measuring Machine

A refined method of carrying out the above measurement is to use a machine as shown in Fig. 4.6. It consists essentially of a micrometer, reading to 0·0002 μm units, and a fiducial indicator, which is free to 'float' across the line of measurement. The measuring head can be raised or lowered on a stiff column, and the work is supported on a heavy base having a lapped surface and is held in position by a centre at the top.

The micrometer is set to a cylindrical standard of known size and with care high orders of accuracy can be maintained.

Fig. 4.6.   Taper measuring machine.
(*By courtesy of the Coventry Gauge & Tool Co. Ltd.*)

4.33 Measurement of Taper Ring Gauges

The procedure here is similar to that used for taper plug gauges, the measurements being made with precision calibrated balls and gauge blocks.

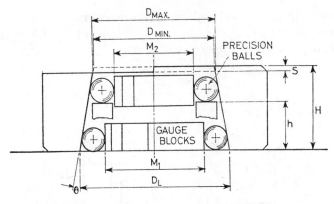

Fig. 4.7. Measurement of a taper ring gauge.

Again it can be seen that

$$\tan \theta = \frac{M_1 - M_2}{2h}$$

But in this case

$$D_L = M_1 + d \left[1 + \cot \frac{90 - \theta}{2} \right] \quad \text{(where } d = \text{ball dia.)}$$

and $D_{\max} = D_L - 2(H - S) \tan \theta$
$$D_{\min} = D_L - 2H \tan \theta$$

The accuracy obtainable by this method depends largely on obtaining the correct 'feel' between the balls and the gauge blocks and this comes only with practice and experience. However, it should be remembered that in making such a measurement M there are four point contacts, all offering a probability of elastic compression, and that a gauge block just too large exerts a great wedging force. The authors feel that the best method is to start with the gauge blocks undersize and to see how much free movement of one of the balls is possible. The ball can be manipulated with a steel knitting needle. As the gauge block sizes are increased in 0·01 mm increments the free movement of the ball is reduced until finally no movement at all is possible. If the arrangement is left to cool and stabilize after handling, a final check can be made on the 'feel' of the balls and gauges, and any necessary size adjustments made.

This method at least provides an objective process in approaching the correct 'feel' and is not simply a method of guessing at a 'nice snug fit' which nobody defines.

4.34 Measurement of Taper Bores

The procedure described in section 4.32 may only be used where the taper bore is of large enough diameter to allow easy access of balls and gauge blocks. For smaller diameter tapers, angular and diametral measurements may be made using the arrangement shown in Fig. 4.8, in which the height at which balls of different diameters seat directly on the taper is measured.

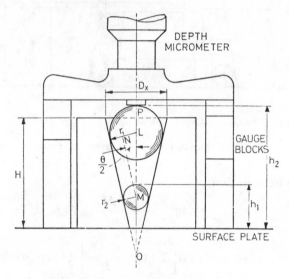

Fig. 4.8. Measurement of a taper bore.

The centre distance LM between the balls will be

$$\text{LM} = h_2 - h_1 - r_1 + r_2$$

$$\sin \frac{\theta}{2} = \frac{\text{NL}}{\text{LM}} = \frac{r_1 - r_2}{h_2 - h_1 - r_1 + r_2}$$

Having thus found the semi-angle of the taper, its uniformity may be determined by taking a further measurement on a ball of such diameter that it rests approximately mid-way along the length of the taper, and repeating the calculation for the new measured values obtained.

To find the diameter D_x at the large end of the taper:

If O is the apex of the taper

$$\frac{\text{OL}}{\text{LM}} = \frac{r_1}{r_1 - r_2}$$

$$\text{and } \text{OP} = \text{OL} + \text{PL}$$

$$\text{and } D_x = 2\left(\text{OP} \tan \frac{\theta}{2}\right)$$

The practical difficulties of this method of measurement are perhaps more severe even than in the previous case, especially if the taper has a small angle. The wedging effect of the balls and the consequent elastic deformation both of the balls and the gauge can cause appreciable errors in the measured values of h_1 and

h_2, to give rise to related errors in the calculated angle. On no account should the balls be dropped into the taper. It is better if they are gently rolled into position with the axis of the taper lying almost horizontal. Again, here is a case where practical experience of such measurement allows one to overcome the difficulties involved.

4.4 THE PRECISION LEVEL

Reference has already been made in Chapter 3 to the Brookes Level Comparator, in which the basis of comparison of the height of two gauge blocks is the sensitivity of a spirit level.

Essentially the spirit level is an angular measuring device in which the bubble always moves to the highest point of a glass tube, the bore of which is ground to a large radius. The sensitivity of a spirit level is governed solely by the radius of the tube or vial containing the liquid, and by the base length of its mount.

Assume a level has graduations on the vial separated by a distance of l and the tube is of radius R.

Now let one end of the tube be raised so that it is moved through an angle δ. If this causes the bubble to move 1 division, then:

$$\delta \text{ radians} = \frac{l}{R}$$

If the graduations are at 2·5 mm intervals and these represent a tilt of 10 sec of arc, then:

$$10 \text{ sec} = 0 \cdot 000 \ 048 \ 5 \text{ radians} = \frac{2 \cdot 5 \text{ mm}}{R}$$

$$R = \frac{2 \cdot 5 \text{ mm}}{0 \cdot 000 \ 048 \ 5} \qquad = 51 \ 500 \text{ mm}$$

or $R = 51 \cdot 5$ m approximately

Using this vial radius then in a base whose length is 250 mm, the height x which one end must be raised for 2·5 mm bubble movement is given by:

$$0 \cdot 000 \ 048 \ 5 \text{ radians} = \frac{x}{250 \text{ mm}}$$

or $x = 0 \cdot 012$ mm

Similarly as the base length is reduced from 250 mm, so the sensitivity is increased. Thus, if the standard base of 250 mm is reduced to 125 mm, then each scale graduation represents 0·006 mm.

The main use of a precision level lies, not in measuring angles, but in measuring straightness and wind in machine tool slideways. This is dealt with more fully in Chapter 6, but briefly, if the level is stepped along the slide in intervals of its own base length, the first position being taken as a datum, height of all other points relative to this datum can be determined.

Although simple in concept and use, it should be remembered that the accuracy of a given level depends on the setting of the vial relative to the base. In most designs provision is made for adjustment of the vial in its mounting. However, in taking precise measurements of level, it is as well to assume that an error in the vial setting does exist and to take two readings along the same line, but in opposite directions. The mean of these readings will indicate the true amount of error in level of the surface.

4.41 The Clinometer

The clinometer is a special case of the application of the spirit level. In this instance the level is mounted in a rotatable body carried in a housing, one face of which forms the base of the instrument. A main use of the instrument is the measurement of the included angle of two adjacent faces of a workpiece. Thus, in use, the instrument base is placed on one face and the rotatable body is adjusted until a zero reading of the bubble is obtained. The angle of rotation necessary to bring this about is then shown on an angular scale moving against an index.

A second reading is taken in a similar manner on the second face of the workpiece, the included angle between the faces being the difference between the first and second readings. Depending upon the type of instrument used, readings direct to 1 min are obtained, and up to a range of movement of 90°.

4.42 Standards for Spirit Levels

Standard of precision, sensitivity, calibration, and condition and accuracy of the locating faces, are dealt with in B.S. 958. For general precision work a sensitivity of 10 sec is most useful, i.e. an angular displacement of 10 sec causes a bubble movement of 1 division on the vial, usually about 2·5 mm.

4.5 OPTICAL INSTRUMENTS FOR ANGULAR MEASUREMENT

For the measurement of small angular differences a group of optical instruments is available, all of which depend in principle on the collimation of a beam of light.

If a point source of light O is placed at the principal focus of a collimating lens it will be projected as a parallel beam of light as in Fig. 4.9 (*a*). If this parallel beam now strikes a plane reflector which is normal to the optical axis, it will be reflected back along its own path and refocused at the source O.

If the plane reflector is now tilted through some small angle δ, the reflected parallel beam will turn through 2δ, and will be brought to a focus at O_1, in the focal plane, a distance x from O.

This effect is shown in Fig. 4.9 (*b*). If the ray passing through the geometric centre of the lens is considered, as it is, unaffected by refraction, it can be seen that:

$$x = 2\delta f \text{ mm}$$

where *f* is the focal length of the lens.

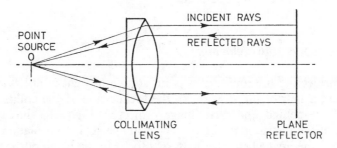

Fig. 4.9(*a*).   Point source of light in focal plane of a collimating lens.

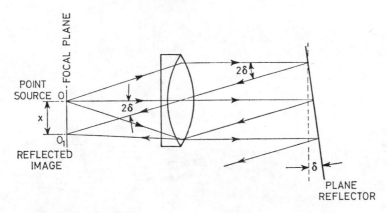

Fig. 4.9(*b*).   Projection of a point source being reflected from an inclined reflector.

There are certain important points about this expression which are not immediately apparent. These are:

(*a*) The distance between the reflector and the lens has no effect on the separation *x* between source and image.

(*b*) For high sensitivity, i.e. a large value of *x* for a small angular deviation, δ, a long focal length is required.

(c) Although the distance of the reflector does not affect the reading x, if, at a given value of δ, it is moved too far back, all of the reflected rays will miss the lens completely, and no image will be formed. Thus, for a wide range of readings, the minimum distance between lens and reflector is essential. This is particularly important where the principle is used in optical comparators (Chapter 3). However, in this application it simply limits the maximum remoteness of the reflector if the full range of readings of the instrument is to be used.

4.51 The Microptic Auto-collimator

The idea of projecting the image of a point source of light is not practical, so in this instrument a pair of target wires in the focal plane of the collimating lens is illuminated from behind and their image projected. Fig. 4.10 shows the optical arrangement of the instrument, the projected image striking a plane reflector and the reflection of the image being brought to a focus in the plane of the target wires.

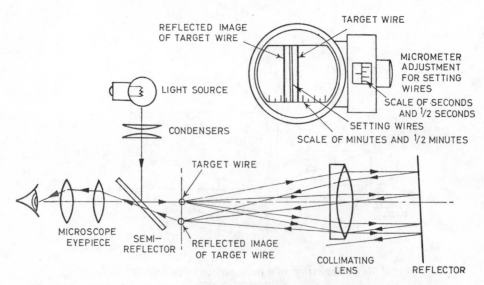

Fig. 4.10. Optical system and view through eyepiece of auto-collimator. A reading is taken by adjusting the micrometer until the setting wires straddle the image.

The wires and their image are viewed simultaneously in an eyepiece, which also contains a pair of adjustable setting wires and a scale. The setting wires are adjusted by a micrometer until they straddle the reflected image (not the target wire). The scale is read to the nearest ½ min, and the micrometer drum which moves the wires ½ min per revolution, is divided into 60 equal parts. Thus 1 division of the micrometer drum represents an angular deflection of the reflector of one $\frac{1}{2}$ sec of arc. With care, and given a rigid mounting for the instrument, repeat readings of 0·2 sec are possible.

The instrument normally has a range of readings of 10 min of arc up to a range of 10 m. It is invaluable in machine tool alignment testing or for any large scale measurement involving small angular deviations. A fuller description of its use for such purposes is given in Chapter 6.

4.52   The Angle Dekkor

In this application of the collimating principle, an illuminated scale is set in the focal plane of the collimating lens outside the field of view of a microscope eye-piece. It is then projected as a parallel beam and strikes a plane reflector below the instrument. It is reflected, and refocused by the lens so that its image is in the field of view of the eyepiece. The image falls, not across a simple datum line, but across a similar fixed scale at right angles to the illuminated image. Thus the reading on the illuminated scale measures angular deviations from one axis at 90° to the optical axis, and the reading on the fixed scale gives the deviation about an axis mutually at right angles to the other two.

This feature enables angular errors in two planes to be dealt with, or more important, to ensure that the reading on a setting master and on the work is the same in one plane, the error being read in the other. Thus induced compound angle errors are avoided.

The optical system and the view in the eyepiece are shown in Fig. 4.11. The physical features simply consist of a lapped flat and reflective base above which the optical details are mounted in a tube on an adjustable bracket.

In use, a master, either a sine bar or a group of combination angle gauges (see section 4.521) is set up on the base plate and the instrument adjusted until a reading on both scales is obtained. It is now replaced by the work, a gauge block to give a good reflective surface being placed on the face to be checked. The gauge block can usefully be held in place with elastic bands. The work is now slowly rotated until the illuminated scale moves across the fixed scale, and is adjusted until the fixed scale reading is the same as on the setting gauge. The error in the work angle is the difference in the two readings on the illuminated scale.

To check the angle between faces which are nominally at 90° to each other, no master is necessary. If gauge blocks are held against both faces and the angle-dekkor adjusted to give a reading it will be found to consist of two mirror images (due to double reflection) of the illuminated scale, superimposed on each other. The misalignment of the readings of these images will be *double* the error in the right angle.

The type of view obtained in the eyepiece when the angle dekkor is used in this manner is shown in Fig. 4.12. Initially it is difficult to read, but with practice the method provides a quick and accurate means of testing right angles.

Although the angle dekkor is not as sensitive as the auto-collimator it is extremely useful for a wide range of angular measurements at short distances.

It therefore finds its application in toolroom-type inspection. Readings direct to 1 min over a range of 50 min may be taken, and by estimation, readings down to about 0·2 min are possible.

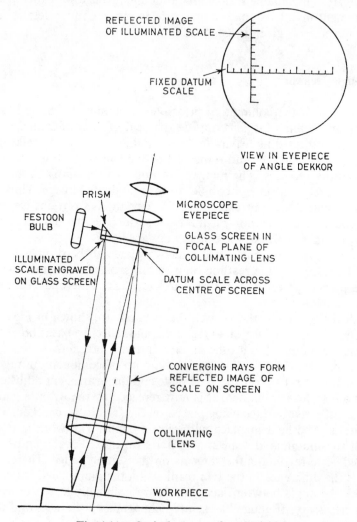

REFLECTED IMAGE
OF ILLUMINATED SCALE

FIXED DATUM
SCALE

VIEW IN EYEPIECE
OF ANGLE DEKKOR

PRISM

MICROSCOPE
EYEPIECE

FESTOON
BULB

GLASS SCREEN IN
FOCAL PLANE OF
COLLIMATING LENS

ILLUMINATED
SCALE ENGRAVED
ON GLASS SCREEN

DATUM SCALE ACROSS
CENTRE OF SCREEN

CONVERGING RAYS FORM
REFLECTED IMAGE OF
SCALE ON SCREEN

COLLIMATING
LENS

WORKPIECE

Fig. 4.11. Optical system of angle dekkor.

4.521 Combination Angle Gauges

It was mentioned in section 4.5 that the angle dekkor, which is essentially an angle comparator, was set to a master, as is any other comparator. It is possible to use a sine bar, but in 1941 a much more convenient form of setting master, the combination angle gauge set, was introduced.

Combination angle gauges are simply blocks, hardened and lapped to precise angles, so that they can be 'wrung' together. Unlike gauge blocks, it must be realized that angular blocks can be added or *subtracted* as in Fig. 4.13.

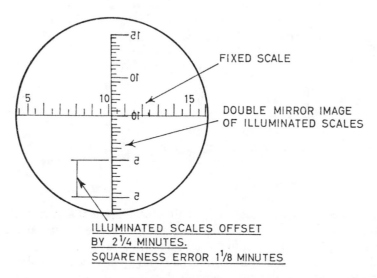

Fig. 4.12.   View in angle dekkor eyepiece when checking squareness direct.

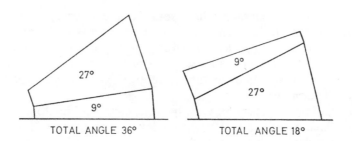

Fig. 4.13.   Addition and subtraction of combination angle gauges.

The values of the angles used are arranged in a modified geometric progression with a common ratio of 3 as shown on the following page.

These thirteen gauges, along with a square block each of whose angles are calibrated, enable any angle between 0° and 90° to be realized in increments of 3 sec. The gauges are used either in combination with each other or may be subtracted from the square block.

The gauges, which are manufactured to the same high standards as gauge blocks, are stabilized, have 'wringing' characteristics, and are calibrated to a high degree of precision. The reflective properties of their lapped surfaces make them particularly suitable for use with collimating type of instruments.

Table of Nominal Values of Combination Angle Gauges

Degrees	Minutes	Decimal Minutes
1	1	0·05
3	3	0·1
9	9	0·3
27	27	0·5
41		

4.53 The Alignment Telescope

The principal use of this instrument is in testing the alignment accuracy of bearings, location surfaces on large assembly fixtures, and work of a similar nature.

It consists of two units, a collimating unit and a focusing telescope, the body of each of which is ground truly cylindrical and to a precise outside diameter. Further, the optical axis and the mechanical axis are coincident. Thus each unit may be fitted directly or by precision bushing into two bearings a considerable distance apart, and sightings taken from the telescope unit to the collimating unit.

The collimating unit contains a light source and condensers, in front of which is placed an angular graticule in the focal plane of the collimating lens. This graticule scale is thus projected as a parallel beam of light. If the telescope is focused at infinity it will bring to a focus the parallel rays and the angular scale is seen against the datum lines in the telescope. Thus angular misalignment in both planes is determined.

The collimating unit also contains, in front of the collimating lens, a second graticule also having two scales at right angles to each other. If the focus of the telescope is now shortened this graticule is seen against the datum lines of the telescope and linear displacements are measured directly. In this case the collimating lens is simply providing even illumination for the displacement graticule, and the angular misalignment graticule cannot be seen because it is so far out of focus. A line diagram of the collimating unit and the view in the telescope eyepiece at both conditions of focus is seen in Fig. 4.14.

It should be remembered that distance has no effect on the angular misalignment readings as these are taken using the collimating principle. However, this is not so with the linear displacement scale. The telescope only magnifies the apparent size of the scale as seen by the eye. Thus, as the distance is increased, the size of the displacement scale is reduced and its 'readability' is also reduced. Thus the accuracy of this reading diminishes with distance.

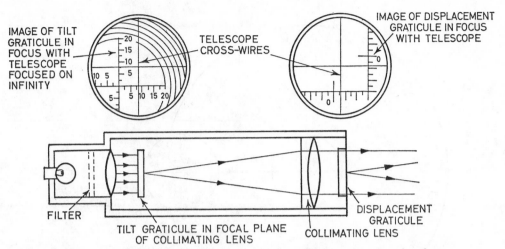

Fig. 4.14. Collimating unit of an alignment telescope showing views in telescope eyepiece with two conditions of focus.

4.6 CIRCULAR DIVISION

Most of the early work in line standards of measurement was carried out by Société Genevoise, a Swiss firm who have since produced most of the highly precise line standards, including the International Prototype Metre and its copies.

Thury, the scientist responsible for this work, realized that the making of high-class instruments requires precisely divided circular scales, and developed a circular dividing machine in 1865.

At this time, there was no method of checking the spacing of the teeth around the table of the dividing machine, and Thury set out to use the linear dividing machine, used for manufacture of line standards, as a form of sine bar to calibrate the dividing machine.

The first divided scale used for circular division was thus made. Since then many applications have been found for such scales in optical rotary tables, optical dividing heads, etc.

4.61 Mounting of Divided Glass Scales

The simplest form of optical divider consists of a glass scale which can be mounted on a mandrel and viewed through a microscope as in Fig. 4.15.

The scale is illuminated from behind and may be read directly and in relation to the scale in the microscope eyepiece.

It is essential with equipment of this type that the divided circle is mounted so that it rotates truly about its own axis. If any eccentricity is present it will induce errors of division. Consider the disc in Fig. 4.16 as having an eccentricity e, i.e. as it rotates the centre of the disc describes a circle of radius e.

83

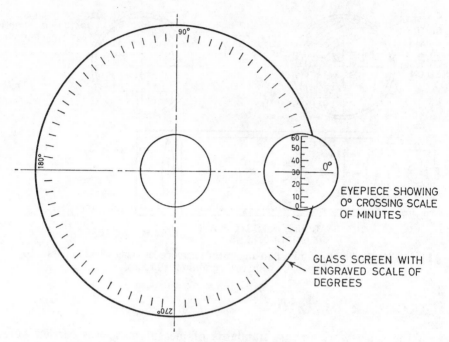

Fig. 4.15.   Principle of optical dividing head or rotary table.

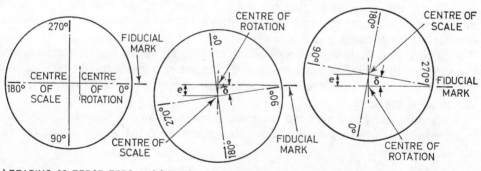

(a) READING 0°, ERROR ZERO   (b) READING 90°, ERROR + δ   (c) READING 270°, ERROR − δ

NOTE: WHEN CENTRES OF ROTATION AND SCALE AND FIDUCIAL
MARK ARE IN LINE THERE IS ZERO ERROR

Fig. 4.16.   Indexing errors due to a divided circle being mounted eccentric by an amount *e*.

In the position shown, the disc will read correctly against the datum.

If turned until it reads 90°, however, its own centre will have moved to the position shown in 4.16 (*b*) and to read 90° it must be turned through an angle of 90 + δ.

Similarly in 4.16 (*c*), to read 270° an equal error of δ occurs. However, in this case the disc is turned through less than 270° to read *that* angle.

Thus the error passes from 0 to $+\delta$, through zero again to $-\delta$, and a graph of the error is of sinusoidal form. The maximum angular error between any two readings is 2δ, and from the diagrams δ radians $=e/R$.

Therefore the maximum angular error, which will occur in this case if the disc is rotated between readings of 90° and 270°, will be $2e/R$, where e is the eccentricity, and R is the radius at which the readings are taken.

Such a divided scale can be used in an optical instrument, or as suggested, mounted on a mandrel and used to enable a workpiece to be rotated through a precise angle.

4.62 Calibrating Circular Divided Scales and Indexing Equipment

The simplest means of calibrating an indexing device, such as a dividing head, is to refer it to a precision polygon.

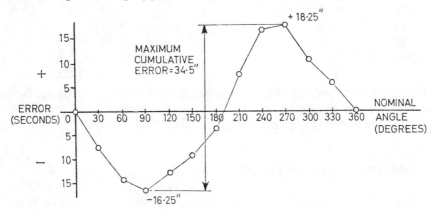

Fig. 4.17. Graph of indexing errors from an indexing device having an eccentric mounting.

This consists of a hardened and stabilized piece of steel whose reflecting faces are accurately lapped so that they are normal to equal divisions of a circle. The largest polygon made has 72 facets at intervals of nominally 5°. Normally a polygon having 12 sides at intervals of 30° is suitable for most work.

The polygon is mounted on the indexing device, and an auto-collimator is set up to give a reflection and reading from any face of the polygon. If the device is now indexed through 30° or $\frac{1}{12}$ of a circle, the reading should be repeated in the auto-collimator. If this is not so, then the difference between the readings is the error in indexings. Similarly, the difference between all readings and the first, are the errors in indexing through the total angle, while the errors between individual indexing motions are obtained by subtraction.

Normally a repeat reading is taken on the original facet of the polygon and any error in repetition is evenly distributed among the separated indexing motions.

The table below is for a set of readings on a milling machine dividing head set for simple indexing, the results being set out in graphical form in Fig. 4.17.

Position	Reading min	Reading sec	Cumulative Error (sec)	Correct for Zero (sec)	Actual Error (sec)
0	4	53	0	0	0
1	4	45	− 8	+0·25	− 7·75
2	4	38	− 15	+0·50	− 14·50
3	4	36	− 17	+0·75	− 16·25
4	4	39	− 14	+1·00	− 13·00
5	4	42	− 11	+1·25	− 9·75
6	4	48	− 5	+1·50	− 3·50
7	4	59	+ 6	+1·75	+ 7·75
8	5	08	+15	+2·00	+17·00
9	5	09	+16	+2·25	+18·25
10	5	01	+ 8	+2·50	+10·50
11	4	56	+ 3	+2·75	+ 5·75
12	4	50	− 3	+3·00	0
(Repeat 0)					

This tabulation presupposes that the polygon is perfectly accurate. None are absolutely accurate of course, but so long as the angular error at each face is known then allowance can be made for this error.

4.63   Calibrating a Precision Polygon

If the errors in a polygon are not known then it can be calibrated using two auto-collimators set up to give reflections from adjacent polygon faces as shown in Fig. 4.18 on p. 87.

If R_1 and R_2 are the readings on auto-collimators 1 and 2 respectively, S is the angle between the normals to faces A and B, and T is the angle between the auto-collimators, then,

$$S + R_1 = T + R_2$$
$$\text{or } S = T + (R_2 - R_1) \qquad \qquad \text{... (1)}$$

If the complete polygon is considered and all values of S, T, and $(R_2 - R_1)$ added then,

$$\Sigma S = \Sigma T + \Sigma (R_2 - R_1)$$

But $\Sigma S = 360°$

$$\therefore 360° = \Sigma T + \Sigma (R_2 - R_1)$$

Dividing through by n, the number of sides of the polygon,

$$\frac{360°}{n} = \frac{\Sigma T}{n} + \frac{\Sigma (R_2 - R_1)}{n}$$

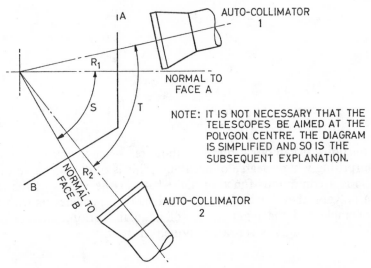

Fig. 4.18.   Arrangement of two auto-collimators set up to measure
the errors in a precision polygon.

Now $360°/n$ is the nominal angle of the polygon and, as T is a constant, the
angle between the auto-collimators, then $\Sigma T/n = T$.

$$\frac{360}{n} = T + \frac{\Sigma(R_2 - R_1)}{n}$$

$$\text{or } T = \frac{360}{n} - \frac{\Sigma(R_2 - R_1)}{n}$$

From a complete series of readings $\Sigma(R_2 - R_1)$ can be established and thus
the value of T determined. This value can be substituted back in equation (1)
for each face and the angle of each face determined.

The calculation is normally set out in tabular form as shown below:

Faces Viewed	R_1		R_2		$(R_2 - R_1)$		Error (sec)
	min	sec	min	sec	min	sec	
0– 45	4	32·5	7	48·7	3	16·2	+2·8
45– 90	5	15·3	8	33·8	3	18·5	+5·1
90–135	2	17·6	5	29·9	3	12·3	−1·1
135–180	4	52·1	8	07·2	3	15·1	+1·7
180–225	5	03·5	8	10·2	3	06·7	−6·7
225–270	4	22·9	7	41·0	3	18·1	+4·7
270–315	1	16·8	4	25·3	3	08·5	−5·9
315– 0	4	18·9	7	30·7	3	11·8	−1·6
Total					25	47·2	

$$T = \frac{360°}{n} - \frac{\Sigma(R_2 - R_1)}{n}$$

$$= 45° - \frac{25' \ 47 \cdot 2''}{8}$$

$$= 45° - 3' \ 13 \cdot 4''$$

$$S = \ T + (R_2 - R_1) = 45° - \frac{\Sigma(R_2 - R_1)}{n} + (R_2 - R_1)$$

$$\text{Error} = \ (R_2 - R_1) - 3' \ 13 \cdot 4''$$

Thus the polygon can be calibrated without reference to any angular standard, and to a surprisingly high degree of accuracy. This is of course based on the fact that a circle is a continuous function, i.e. whatever the values of the individual angles of a polygon they total *exactly* 360°. This then provides us with a natural standard of angle and the maintenance of angular working standards such as combination angle gauges is considerably helped by it.

4.7 MEASUREMENT OF SQUARENESS

There is one particular angle which is probably more important than any other, i.e. 90°. Over the years its importance has been emphasized by the fact that special names—square, normal, and more particularly the right angle—have been assigned to it.

If a lathe cross-slide does not move at 90° to the spindle axis then a flat face will not be produced during a facing operation. If a depth micrometer spindle is not square to the locating face an incorrect measurement, involving a cosine error, will be incurred. If the blade of a fitter's square is not normal to the stock then mating parts will not fit together as they should. In fact if a right angle was unattainable to within a close degree of accuracy it is doubtful whether the achievements of modern science would have reached their present state of advancement.

Fortunately an angle of 90° lends itself to accurate measurement, as in most of the techniques used a doubling effect is possible. An example of this has already been given in section 4.52 as a particular use of the angle dekkor. Often it is not enough to test a workpiece against an engineer's square. This simply shows whether it is a right angle, as near as can be judged by eye, or not. More often the amount of error, either as an angle or as a linear measurement over a given length of face, is required. In any case each result can be readily converted to the other form.

4.71 Contact Methods of Measuring Squareness

Consider a block whose opposite faces are supposed to be parallel, and whose adjacent faces are nominally at right angles. The parallelism of opposite faces can be readily checked by means of a micrometer or comparator depending on the degree of accuracy required, and parallelism can be produced by a number of means, surface grinding, spot grinding, etc., again depending on the degree of accuracy required.

Let us assume that the block is accurately parallel but out of square. If a parallel strip of known accuracy is clamped against an angle plate in some attitude near the vertical, then gauge blocks of differing size can be trapped between the block and the parallel strip as in Fig. 4.19 (*a*).

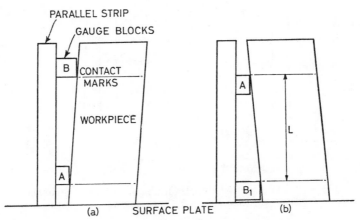

Fig. 4.19. Squareness test using gauge blocks. Squareness error $= \frac{1}{2}(B - B_1)$ over length L.

The contact position of the slip gauge is noted and transferred across the block, which is then turned through 180° and gauge block A is now trapped at the top contact position.

If the block is square a slip gauge equal in size to B should just fit at the lower contact position. If it does not, but a gauge block of size B_1 just contacts correctly then the difference $B - B_1$ is *double* the error in squareness in the workpiece over the length between the contact marks.

A similar, but more convenient method, is to use a measuring instrument designed for this purpose incorporating a dial test indicator or small comparator head as shown in Fig. 4.20.

In this case the difference in the dial gauge readings is *double* the squareness over a length of work equal to the centre distance between the fixed contact and the dial gauge.

4.711   *Correction of Squareness Error*

It is of little value knowing the error in squareness of a workpiece without being able to correct it. Fortunately this is not difficult.

Referring again to Fig. 4.20 let us assume that the difference in dial gauge readings is 0·12 mm, i.e. the block is out of square by 0·06 mm over the length of surface of the contact tip and dial gauge centre distance. Correcting this value for the complete length of surface let us say that the block is out of square by 0·075 mm. To correct this, point A must have 0·075 mm removed from it, nothing being taken off D. Similarly C must be brought 0·075 mm towards D, no metal being removed at B.

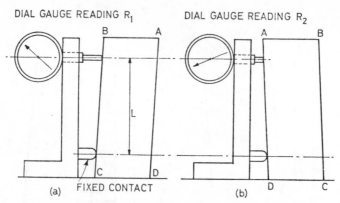

DIAL GAUGE READING R_1　　　　DIAL GAUGE READING R_2

(a)　FIXED CONTACT　　　(b)

Fig. 4.20.　Dial gauge fixture for testing squareness. Squareness
error $= \frac{1}{2}(R_1 - R_2)$ over length L.

If the block is now set up on a surface grinder, face AD uppermost, a 0·075 mm cut can be taken across this face to within about 2·5 mm of D as shown in Fig. 4.21 (*a*).

The block is now turned over as in 4.21 (*b*) and a cut taken to clean up face BC, which has now been corrected by the required amount and is therefore square with reference to AB and CD.

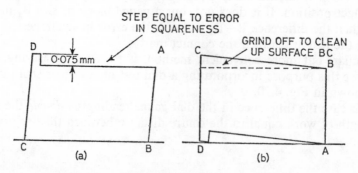

STEP EQUAL TO ERROR
IN SQUARENESS

GRIND OFF TO CLEAN
UP SURFACE BC

0·075 mm

(a)　　　(b)

Fig. 4.21.　Method of correcting squareness.

It now only remains to grind AD parallel to BC again for all four faces to be 'true', i.e. adjacent faces square, and opposite faces parallel.

4.72 Optical Methods of Checking Squareness

A similar principle is involved in a simple fixture for use in conjunction with an auto-collimator or angle dekkor. As with so many simple, yet extremely effective, metrological devices, credit must be given to the National Physical Laboratory for its design.

This squareness tester consists simply of a bar pivoted at a convenient point, carrying a pair of hardened and ground steel cylinders of precisely the same size. The bar also has mounted upon it a plane reflector, and the instrument is set up on a flat reference plane of a suitable degree of accuracy, usually a Grade A surface plane.

The arrangement is shown in Fig. 4.22.

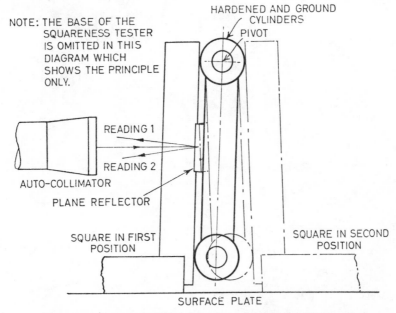

Fig. 4.22.   Squareness test using an auto-collimator. Squareness error $= \frac{1}{2}(R_1 - R_2)$.

A reading is taken with the auto-collimator with the square in the first position shown. The square is then moved to position 2, and the auto-collimator reading again noted. The angular error in squareness is *half* the difference in the two readings.

It should be noted that apart from errors in reading the auto-collimator, and human failings such as not properly cleaning contact surfaces, the only possibility of error using this method is due to cylinders of unequal diameter. This error can be eliminated by exchanging the cylinders and repeating the process.

The actual error in squareness is the mean of the two results.

It is of interest to note that all of the methods of testing squareness referred to in this section are fundamental and do not refer to any standard of angle. They depend solely on the peculiar geometric properties of this important 'right' angle. Other methods of testing are available but have purposely been ignored since, due to the fact that they depend on the accuracy of ancillary equipment, errors may be introduced. For instance in Chapter 6, dealing with machine tool metrology, reference is made to the pentagonal prism, or 'optical square' for checking the squareness of machine tool slideways. It should be noted that this method depends on the accuracy of the prism and if the angle between its reflecting faces is not 45° the error incurred will be double the error in the prism.

CHAPTER 5

Limits and Limit Gauges

5.1 GENERAL

THE correct and prolonged functioning of most manufactured articles depends on the correct size relationships between the various components of the assembly. This means that the parts must fit together in a certain way, e.g. if a shaft is to rotate in a hole there must be enough clearance between the shaft and the hole to allow an oil film to be maintained, but not so much clearance that excessive radial float is allowed. Similarly if the shaft is to be held tightly in the hole there must be enough interference between the shaft and the hole to ensure that the forces of elastic compression grip tightly and do not allow movement. However, the interference must not be excessive or the member containing the hole may split.

Ideally any such condition could be obtained by specifying a definite size for the hole and for the shaft, but this unfortunately is not possible for two very good reasons.

(a) It is impossible to make a part to a specified definite size except by remote chance.

(b) If by chance a part is made exactly to the size required, it is impossible to measure it accurately enough to prove it.

If one examines B.S. 4311 dealing with gauge blocks it is found that grade 00 gauge blocks of up to 25 mm in length are accurate to within 0·05 μm, but they *are not exact.*

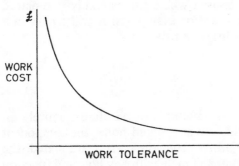

Fig. 5.1. Graph showing type of relationships between tolerance and cost—accuracy is expensive.

Further, it should be noted that as the required degree of precision for a component increases so does its cost. This is not a linear relationship as it costs little, if any, more to make a part to within 0·25 mm of its nominal size, than to make it within 0·50 mm. However, it costs much more to make a part to within 0·002 mm than to within 0·01 mm of nominal size. Thus the form of a graph of accuracy against cost is as shown in Fig. 5.1, the curve being asymptotic to the cost axis, i.e. absolute accuracy costs an infinite amount.

If then a part cannot be made exactly to its nominal size how can two parts be made to fit together in the required manner?

There are three possible answers to this question, each method having its place in industry.

5.11 Making to Suit

This technique simply requires one part to be made to its nominal size as accurately as is economically reasonable. The other part is then machined away a small amount at a time and the parts offered until they fit in the required manner. This method may be used for 'one off' jobs, toolroom work and so on, where both parts will be replaced at once. As such it is of little concern in this book.

5.12 Selective Assembly

It is sometimes found that it is not economic to manufacture parts to the required high degree of accuracy for their correct functioning. Instead they are made in an economic manner, measured to the required accuracy and graded, or sorted into groups each of which contains parts of the same size to within close limits. They are then assembled with mating parts which have been similarly graded.

A good example of this system occurs in ball-bearing manufacture. A ball bearing consists essentially of an inner ring, and an outer ring, separated by steel balls. Both types of ring and the balls are graded automatically and when assembled the following conditions can be allowed:

(*a*) Large balls are assembled into small inner and larger outer rings.

(*b*) Medium balls are assembled into medium inner and outer rings, large inner and outer rings, or small inner and outer rings.

(*c*) Small balls are assembled into large inner and small outer rings.

Usually a selective assembly system is used where the assembly is replaced as a unit rather than replacing separate parts. For example, if a ball cracks in a bearing the whole bearing is replaced, not just one ball.

5.13 Systems of Limits and Fits

Repetitive production of components and assemblies is based almost entirely on interchangeable manufacture. Again considering shafts and holes, the component sizes are specified and the allowable deviations from these sizes are stated, the allowable deviations being such that any shaft will mate with any hole and function correctly for the designed life of the assembly.

There are three basic types of fit obtainable using this method. These are:

(*a*) *Interference fit*. The minimum permitted diameter of the shaft is larger than the maximum allowable diameter of the hole.

(*b*) *Transition fit*. The diameter of the largest allowable hole is greater than that of the smallest shaft, but the smallest hole is smaller than the largest shaft.

(*c*) *Clearance fit*. The largest permitted shaft diameter is smaller than the diameter of the smallest hole.

These conditions are shown in Fig. 5.2.

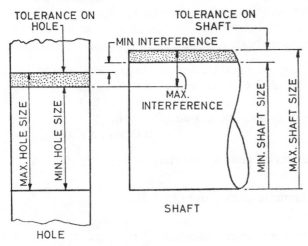

(a) INTERFERENCE FIT

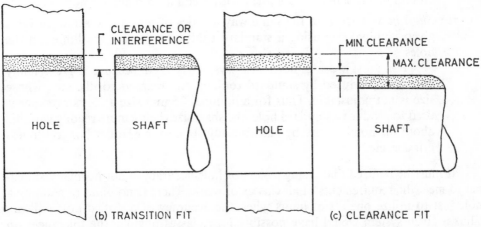

Fig. 5.2. Possible size relationships between a hole and a shaft.

The diagram in Fig. 5.2 also illustrates the terms used in this type of work definitions of these terms being as follows:

(a) *Limits of size*. These are the maximum and minimum sizes allowed on a given component.

(b) *Tolerance* is the maximum size variation which will be tolerated on a given component.

(c) *Allowance* is the size difference between the limiting conditions of size on the two components. It should be noted that a positive allowance produces a clearance fit, and a negative allowance produces interference.

A given manufacturing organization will require a number of different types of fit, ranging from tight drive fits requiring shrinkage or hydraulic pressure for assembly, through keying fits for locations which can be assembled with light mechanical pressure, to running fits and clearance fits. Such a series of fits can be obtained using two distinct policies or basic methods.

(a) *Hole basis system*. For a given nominal size the limits on the hole are kept constant and a series of fits are obtained by varying the limits on the shafts.

Thus assuming a hole of dimension 25 mm + 0·02 mm
$$- 0·000 \text{ mm, a shaft of}$$

(i) 25 mm + 0·08 mm diameter gives an interference fit;
+ 0·04 mm

(ii) 25 mm + 0·02 mm diameter gives a transition fit;
− 0·000 mm

(iii) 25 mm − 0·02 mm diameter gives a clearance fit;
− 0·05 mm

and all of these fits are obtained with a common diameter hole.

(b) *Shaft basis system*. In the same way a series of fits can be arranged for a given nominal size using a standard shaft and varying the limits on the holes.

For most work a hole basis system is used because a great many holes are produced by standard tooling, e.g. reamers, drills, etc., whose size is not adjustable. Thus for a nominal 25 mm size a 25 mm reamer is used to produce a standard hole, the shaft sizes being more readily variable about the nominal size by machine adjustments, e.g. roller box, centreless grinder, etc.

It must be realized that a single class of hole accuracy is not normally sufficient to meet the requirements of all classes of work. There is no point in reaming a hole just to utilize one set of limits when the accuracy of a drilled hole will do. Thus a given system would have possibly four classes of hole, the tolerances on which allow for four different production methods. It might also require, say, nine different classes of shafts to give nine different types of fit when associated with a given class of hole.

Further it must be realized that larger sizes require greater tolerances, and consideration must be given to this. A working system of limits and fits is therefore fairly complex if a manufacturing organization is setting one up from first principles. Fortunately there are standard systems already available.

5.131 The Newall System of Limits and Fits

This system is very popular because of its simplicity. It is a hole basis system providing two classes of holes for different degrees of precision of production. Limits are also given for six classes of shaft; two interference, one transition, and three running fits. These limits are specified for a wide range of sizes of work, the whole system being set out in the form of a table which makes it very convenient to use.

5.132 The British Standard System (B.S. 4500: 1969)

B.S. 4500: 1969 is a comprehensive system designed to cater for all classes of work from instruments and gauges to large heavy engineering. At first it appears to be an extremely complex system until it is realized that no one organization should attempt to use all of it. Instead a company selects from within the standard system a 'sub-system' to suit its own requirements and manufacturing techniques.

The system provides 28 *types* of hole designated by a capital letter A, B, C, D, ..., etc. and 28 types of shaft designated by a lower case letter, a, b, c, d, ..., etc. These letters define the position of the tolerance zone relative to the nominal size. For instance, all class H holes have limits of $^{+x}_{-o}$. Similarly class h shafts have limits of $^{+o}_{-x}$ and so on.

To each of these types of hole or shaft may be applied 18 grades of tolerance designated by numbers, so that a hole may be designated H7, a shaft j4, and a fit between the two is specified H7/j4.

Thus an organization requiring a hole basis system would select from B.S. 4500 a series of grades of hole all of one type to suit its work. Then a series of shafts to give the required number and types of fit would be selected. A typical hole basis system would be:

Holes: H4   H7   H9   H11

Shafts: u6,   s6,   p6   m6,   k6   h7, g6, f7, e7   (used in

              ‿‿‿‿‿‿‿‿‿   ‿‿‿‿‿‿‿‿   ‿‿‿‿‿‿‿‿   association

              Interference   Transition   Clearance   with H7 hole).

The requirements of many organizations can be covered by a very small range of fits and to meet these requirements the British Standards Institution publishes data sheets 4500A and 4500B, being selected fits of hole basis and shaft basis respectively; each provides for six clearance fits, three transition fits and four interference fits.

5.2 LIMIT GAUGES

Adoption of a system of limits and fits logically leads to the use of limit gauges, with which no attempt is made to determine the size of a workpiece—they are simply used to find whether the component is within the specified limits of size or not. The simplest form of limit gauges are those used for inspecting holes or shafts.

Consider first a hole on which the limits on diameter are specified. It would appear that quite simply the 'GO' gauge is a cylinder whose diameter is equal to the minimum hole size, and that the 'NOT GO' gauge is a similar cylinder equal in diameter to the maximum hole size. Unfortunately it is not as simple as this, for the same reason that limits of size are required for the work; nothing can be made to an exact size and this includes gauges. Thus the gauge maker requires a tolerance to which he may work, and the positioning of this gauge tolerance relative to the nominal gauge size requires a policy decision. For instance, if the gauge tolerance increases the size of a 'GO' plug gauge, and decreases the sizes of the 'NOT GO' end, the gauge will tend to reject good work which is near the upper or lower size limits.

Similarly if the gauge tolerance increases the size of the 'NOT GO' plug gauge and decreases the size of the 'GO' end then the gauge will tend to accept work which is just outside the specified limits.

It follows that a number of questions must be answered in designing a simple limit gauge:

(*a*) What magnitude of tolerance shall be applied to the gauge?
(*b*) How shall the tolerance zones for the gauge be disposed relative to the tolerance zones for the work?
(*c*) What allowance shall be made for the gauge to wear?

These considerations are all dealt with in B.S. 4500 Part 2: 1974*—Inspection of Plain Workpieces

5.21 Gauge Tolerances

As a general guide in any measuring situation the accuracy of a piece of measuring equipment needs to be ten times as good as the tolerance on the work it is designed to measure. This means that the tolerance on a gauge is usually approximately 10% of the tolerance on the work to the nearest 0·001 mm unit.

A detailed examination of B.S. 4500 Part 1 shows that the tolerance grades (see p. 97) are arranged in a series of preferred numbers in groups of five (R5 series) so that the magnitude of any tolerance grade is ten times that of its equivalent in the previous group, thus:

* This standard supersedes B.S. 969 but that standard will remain in being for an interim period to cater for the enormous number of gauges still in use which were manufactured to it.

IT 0	IT1	IT $6 = 10 \times$ IT1	IT11 $= 10 \times$ IT 6	IT16 $= 10 \times$ IT11
IT01	IT2	IT $7 = 10 \times$ IT2	IT12 $= 10 \times$ IT 7	
	IT3	IT $8 = 10 \times$ IT3	IT13 $= 10 \times$ IT 8	
	IT4	IT $9 = 10 \times$ IT4	IT14 $= 10 \times$ IT 9	
	IT5	IT10 $= 10 \times$ IT5	IT15 $= 10 \times$ IT10	

It follows that if a limit gauge is required for a hole specified as H11 then the gauge tolerance will be grade IT6 and so on. This is not strictly followed in B.S. 4500 Part 2 but the principle of relating gauge tolerance to work tolerances is set out in table 1 of that standard. It is of interest to note that the table not only specifies the grade of tolerance to be used for a particular type of gauge but also the tolerance grade to be used for the gauge form. Further, the magnitude of the tolerances vary with the type of gauge according to the difficulty of manufacture.

5.22   Disposition of Gauge Tolerance

Having determined the tolerance magnitude for the gauge it must now be positioned relative to the work limits so that it does not cause the gauge to accept defective work but at the same time does not tend to deprive the production department of an excessive proportion of the work tolerance. In B.S. 4500 Part 2 the disposition of the gauge tolerances is as follows:

(*a*) The tolerance on the 'GO' gauge is bilaterally disposed about a line set a distance 'z' (or z_1 for gauging shafts) within the tolerance zone for the work from the maximum metal limit.

(*b*) The tolerance on the 'NOT GO' gauge is bilaterally disposed about the minimum metal limit for the work.

5.23   Allowance for Wear

Wear allowances are normally only applied to 'GO' gauges. A 'NOT GO' gauge should rarely be fully engaged with the work-piece and should therefore suffer little wear. The allowance for wear on new 'GO' gauges is catered for by placing the tolerance zone for the gauge within the tolerance zone for the work and it therefore remains to define how much wear may be allowed to take place before the gauge should be withdrawn from use and replaced. This is specified by a line positioned a distance 'y' (or y_1) from the maximum metal limit. The value for 'y' and 'y_1' is often zero but where a value is assigned the line is then positioned outside the tolerance zone for the work.

These tolerance dispositions and allowances for wear are shown in Figs 5.3 and 5.4. The limiting sizes for the design and measurement of gauges may be determined by finding 'z' (or z_1), the gauge tolerance and 'y' or (y_1) from B.S. 4500 Part 2 and applying these values as shown in the diagrams.

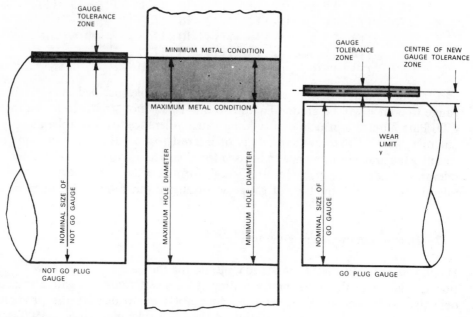

Fig. 5.3.   Disposition of tolerances on plain plug gauges.

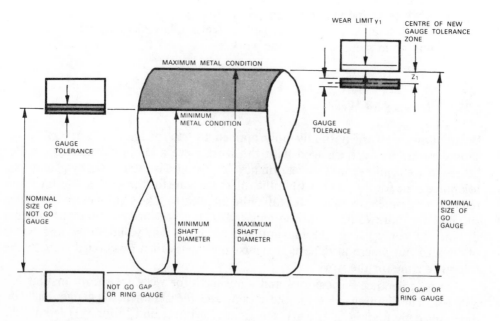

Fig. 5.4.   Disposition of tolerances on plain gap and ring gauges.

For gauges up to and including 180 mm in size this procedure is simplified in tables 3 and 4 of B.S. 4500 Part 2. These tables give the gauge limits and wear limits directly relative to the work limits which must themselves still be determined from B.S. 4500 Part 1.

Thus the dimensions of simple gauges may be dealt with. However, in any design problem, the geometric form of the component, in this case a gauge, must be considered, and under conditions in which the diametral tolerance is small, this assumes considerable significance.

5.3 TAYLOR'S THEORY OF GAUGING

This theory is the key to the design of limit gauges, and defines the function, and hence the form, of most limit gauges. It states:

'The "GO" gauge checks the maximum metal condition and should check as many dimensions as possible.

'The "NOT GO" gauge checks the minimum metal condition and should only check one dimension.'

Thus a separate 'NOT GO' gauge is required for each individual dimension.

Consider a system of limit gauges for a rectangular hole, as shown in Fig. 5.5.

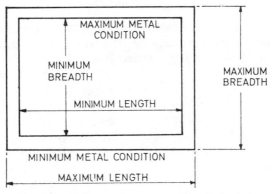

Fig. 5.5. Tolerance zone on a rectangular hole.

The 'GO' gauge is used to ensure that the maximum metal condition is not exceeded and that metal does not encroach into the minimum allowable hole space. It should therefore be made to the maximum allowable metal condition dimensions, due allowance being made for wear and the gauge tolerance as in section 5.2.

Now consider the 'NOT GO' gauge. If this was made to gauge both dimensions of the minimum metal conditions (maximum hole size) a condition would arise where the breadth of the hole is within the specified limits, but the length is over-size, as in Fig. 5.6.

Such a gauge will not enter the hole and therefore the work is accepted although the length is outside the specified limits.

Had separate 'NOT GO' gauges been used for the two dimensions the breadth gauge would have accepted the work but it would have been rejected by the separate length gauge.

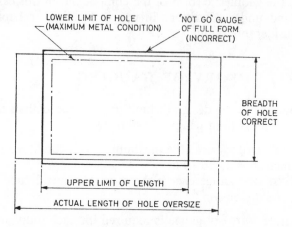

Fig. 5.6. Rectangular hole oversize in one direction.
A full-form 'NOT GO' gauge will not reject such a hole.

This principle should be applied to all systems of limit gauges, and where possible this is done. In the simplest case of a 'GO–NOT GO' plug gauge it would not appear to be so, but to a large extent it is. The 'NOT GO' gauge is always relatively short, and approximately equal in length to the hole diameter. The 'GO' gauge should where possible be equal in length to about three or four diameters. In addition to more readily distinguishing between 'NOT GO' and 'GO' ends of the gauges, the length of the 'GO' gauge ensures that the maximum metal condition is not exceeded due to geometric errors in the work, e.g. straightness as shown in Fig. 5.7.

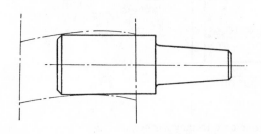

Fig. 5.7. Length of 'GO' plug gauge prevents it entering a non-straight hole.

A similar situation occurs when gauging shafts. Ideally a full form 'GO' gauge, i.e. a ring gauge of reasonable length in relation to its diameter, should be used, in conjunction with a 'NOT GO' gap gauge. In practice both 'GO' and 'NOT GO' gap gauges are frequently used, but it is advisable to supplement these with a 'GO' ring gauge, to be used at frequent intervals. This is particularly true in the case of centreless ground work which is liable to a condition known as lobing.

The simplest lobed condition has three lobes and is based on an equilateral triangle as shown in Fig. 5.8, but it may occur with any *odd* number of lobes.

From each corner of the triangle let radii of *r* and *R* be struck, the large radii blending with the small radii struck from the other two angles.

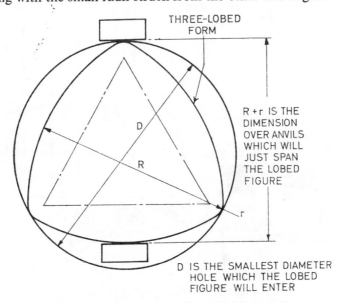

Fig. 5.8.   Effect of lobing on cylindrical work.

The dimension of any such configuration, if measured between a pair of parallel planes, will be $(R+r)$, but the smallest hole that such a form will enter will be much larger as shown by the outer circle. Therefore a gap gauge will accept such work and a diametral measurement with a micrometer will confirm that the work 'diameter' is apparently within limits. However, a 'GO' ring gauge will rightly reject the workpiece. That such a condition is present can be confirmed by setting the component in a vee block and rotating it under a dial gauge or comparator.

5.31   Te-bo Gauges

These are a particular type of plug gauge manufactured by the S.K.F. Ball Bearing Co. Ltd., in Sweden. They are mainly concerned with larger diameters than are gauged by normal plug gauges, and are ground to a spherical diameter equal to the 'GO' limit on the work, due allowance being made for the gaugemaker's tolerance and wear allowance.

A small circular area of the gauge is raised, by electro-plating, an amount equal to the work tolerance. Such a gauge is shown in Fig. 5.9.

It is inserted into the hole by tilting the gauge forward so that the high portion clears the metal. This checks the low limit of size. The gauge is then rocked back until the high point prevents further movement and thus the maximum diameter is tested. If the raised portion passes through the hole then the work is rejected as the hole is oversize.

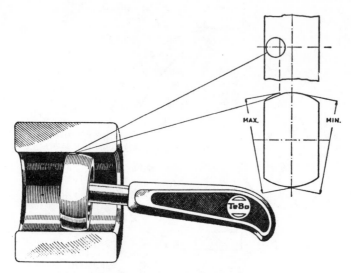

Fig. 5.9. Te-bo-type plug gauge.
(*Courtesy of the S.K.F. Ball Bearing Co. Ltd.*)

It is of interest to note that the 'NOT GO' gauge conforms to Taylor's principle, and only gauges one diameter at a time, i.e. it can check ovality. However, the 'GO' gauge, unlike a normal plug gauge, cannot reject a hole whose diameter is within the allowance limits, but whose lack of straightness causes the maximum metal condition to be exceeded.

5.4 SCREW THREADS

The allowable errors in commercial screw threads are quoted in diameter only, i.e. for a thread of a given pitch and form, tolerances on pitch and flank angle are not given, but are included in the tolerance allowed on effective diameter. Briefly the reason for this is that errors in pitch and flank angle bring about a virtual increase in the effective diameter as noted in Chapter 8. Similar errors on an internal thread bring about a virtual decrease in effective diameter.

The effect of these errors is more fully dealt with in Chapter 8, dealing with screw thread measurements, but the fact that a pitch error, or a flank angle error, can cause an apparent change in effective diameter is most important in any consideration of the design of screw limit gauge, and calls for the application of Taylor's principle.

5.41 Limit Gauges for Screw Threads

Consider first a 'GO–NOT GO' type screw gap gauge in which both 'GO' and 'NOT GO' gauges are made to the full form of the thread, and both of which engage the full length of the thread. It must be emphasized here that such a gauge is *incorrectly* designed, but a consideration of its faults will lead to an explanation of how these faults can be corrected by correctly designing the gauges.

If such a gauge is to be used for inspecting work threaded by means of a self-opening die head, as it frequently is, it will also be of assistance to the machine setter in the initial setting of the die head. Consider in this case that the die head is causing a pitch error.

If the diameter is correct the work will not enter the 'GO' gauge *because of the pitch error*, although it appears that the diameter is too large. The die head would then be adjusted to reduce the thread diameter until it enters the 'GO' gauge, but not the 'NOT GO' gauge. The work thus appears correct, not because the diameter is correct, it is in fact undersize, but because of the diametral compensation by the pitch error.

This is an outstanding example of gauges not conforming to Taylor's principle. If they are redesigned according to this principle, the gauges required would be as follows:

(*a*) 'GO' gauge—Full form and full length of thread to the maximum diameter of the thread.

(*b*) 'NOT GO' gauges—(i) A gauge for the major diameter.
(ii) A separate gauge for the effective diameter and one which is not influenced by pitch errors.

Ideally the 'GO' gauge should be a full form ring gauge, but these are cumbersome in use, and in practice a full form gap gauge is used, a ring gauge being kept for the periodic checks. The plain 'NOT GO' gap gauge is often omitted so that a simple 'GO–NOT GO' screw gap gauge can be used.

To prevent the 'NOT GO' gauge being influenced by pitch errors, it gauges on a short length of thread only, and checks effective diameter only, by being cut away at its crests and roots. Thus the anvils of a 'NOT GO' screw gap gauge appear as shown in Fig. 5.10.

Consider now the workpiece having a pitch error, being gauged by a gap gauge having anvils of this type. The diameter is reduced until the thread passes the 'GO' gauge. When offered to the 'NOT GO' anvils it enters and passes them because the diameter is undersize, and the gauged length of thread is not long enough for the pitch error to be seriously effective.

Similarly, for gauging internal threads, an ideal set of plug gauges would consist of

(*a*) 'GO' gauge—Full form and full length of thread.

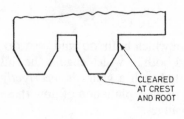

CLEARED
AT CREST
AND ROOT

Fig. 5.10.   Form of anvils on a
'NO GO' screw gap gauge.

(*b*) 'NOT GO' gauge—Cleared at its crest and roots and being only two or three threads in length.

(*c*) 'NOT GO' gauge—Plain plug 'NOT GO' gauge for the minor diameter.

In practice the 'NOT GO' thread gauge is frequently omitted, leaving a simple double-ended plug gauge, and the 'GO' end being the full form, full length of thread gauge, and the 'NOT GO' end being a plain plug gauge to ensure that the minor diameter is not undersize.

5.411 Gauge Tolerances for Thread Limit Gauges

The allowable errors on a limit gauge for a screw thread are tabulated and explained in B.S. 919 Parts 1, 2 and 3; part 3 dealing with gauges for ISO metric threads. It is this part of B.S. 919 which will be considered here. The tolerances for pitch and flank angles are given direct as linear and angular tolerances, unlike past standards where these tolerances were expressed as allowable changes in the effective diameter of the gauge produced by them. It is interesting to note that the tolerances on flank angles increase as the pitch *decreases*, the short flanks of fine pitch threads being much more difficult to control in manufacture and to measure than the longer flanks of coarse pitch threads.

It must also be noted that the limits for the diameters of the gauges are tabulated against the tolerance on the effective diameter of the *product thread* it is to gauge. Thus in order to determine the specification for the gauge the procedure is as follows:

(*a*) From B.S. 3643 find the limits on the major, minor and effective (pitch) diameter of the thread to be gauged.

(*b*) From B.S. 919 find the tolerances or allowable errors on the pitch and flank angles of the gauge.

(*c*) Knowing the tolerance on the effective diameter of the product thread (see (*a*) above) find, from B.S. 919 the limits on the major, minor and effective diameters for the gauge.

(*d*) Knowing the basic gauge sizes (see (*a*) above) apply the gauge limits to these basic sizes to determine the limiting values for each of the gauge diameters.

Considerable care is necessary in reading the tables in B.S. 919 as each table caters for different features of a number of different types of screw thread gauges. Problems may also be caused in finding the limits for the minor diameter of a screw plug gauge. Reference to the diagrams in B.S. 919 reveals the way in which the roots of such gauges may be cleared and the size found is the maximum size, no tolerance being given.

5.42 Reference Gauges for Screw Threads

B.S. 919 also defines tolerance zones, and their magnitude for highly accurate reference screw plug gauges used for setting and testing screw gap gauges, and small screw ring gauges. A screw ring gauge for a very small diameter thread is almost impossible to measure to determine the accuracy of its individual elements. It is therefore checked by using limit plug gauges whose size is accurately made to the high and low limits of the ring gauge itself.

Similar gauges are used to inspect and set screw gap gauges, which are almost impossible to measure by contact methods.

5.5 TAPER LIMIT GAUGES

It is an anomaly that, as far as production inspection is concerned, taper limit gauges do not check the angle of taper of the work. The main purpose of such gauges is to ensure that the diameter at a particular point on the taper is within the specified limits. This is achieved by grinding a step in one end of the gauge so that the diameters at the top and bottom of the step are the limits on the diameter of the taper, and the thumbnail is passed over the gauge to determine by feel that the correct relationship exists between the work and the gauge as shown in Fig. 5.11.

Similarly a taper plug gauge is stepped at the large end.

It is unfortunate that the step on a taper ring gauge is always on the small diameter, and on a taper plug gauge it is usually on the large diameter, since the simplest diameter to determine is the large diameter of a taper ring gauge and the small diameter of a taper plug gauge. These measurements are fully explained in Chapter 4. However, this means that during the manufacture of such a gauge it is not possible to measure the required sizes directly.

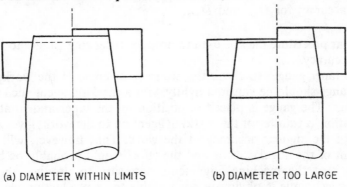

(a) DIAMETER WITHIN LIMITS   (b) DIAMETER TOO LARGE

Fig. 5.11. Method of using a taper limit gauge.

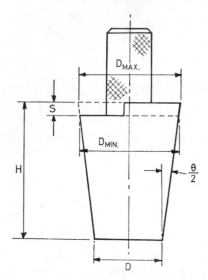

Fig. 5.12. Taper plug gauge.

Consider a taper plug gauge. If the diameter D is determined, the diameters D_{min} and D_{max} are functions of the height of gauge and the angle of taper.

It is shown in Chapter 4 that

$$D_{max} = D + 2H \tan \frac{\theta}{2}$$

and $D_{min} = D + 2(H - S) \tan \frac{\theta}{2}$

Further, the angle of the work to be gauged will have a tolerance, and so will the gauge. If the angular tolerance on the work is say, ± 5 sec, the tolerance on the gauge should be of the order of 1 sec (i.e. 10% of the work tolerance), or $\pm\frac{1}{2}$ sec.

If D is correct, then over a height H, this angular gauge tolerance will make a significant difference to the diameters D_{max} and D_{min}.

It is suggested that the following method of manufacturing these gauges may be used.

(*a*) Make the gauge overlength and to an angle as close as possible to the nominal work angle.

(*b*) Remove the gauge from the machine and accurately determine diameter D and the angle $\theta/2$.

(*c*) Calculate the height H which, based on the above measurements, will give the correct diameter D_{max}. Similarly calculate the height of the step S which will give the correct diameter D_{min}.

(*d*) Grind the length to the correct values of H and S to the required degree of accuracy for D_{max} and D_{min}.

A similar procedure may be used to produce taper ring gauges to the required degree of accuracy.

To use taper gauges to check the accuracy of angle of the workpiece a line along the gauge should be smeared lightly with a marking agent such as prussian blue or rouge. The gauge is placed in position on the taper and rotated slightly. The registration, a transfer of the marking agent on to the work, gives a subjective indication of the angular accuracy of the work. This, however, will not give a measurement of the angular error and the effort would probably be better spent in ensuring that the machine alignments necessary to produce the taper are correct: this technique is obviously not suitable for a production check and can only be used for setting purposes and as a spot check during production.

5.6 HOLE DEPTH GAUGES

Limit gauges for inspecting the depth of holes are also 'thumbnail gauges'. The gauge is normally cylindrical and is made a close sliding fit in a sleeve, to enable it to be removed from the hole, and to give sharp edges against which the step can be felt.

Such a gauge is shown in Fig. 5.13. It must be remembered that not only are tolerances required on the length of the gauge and the step, but also on the thickness of the sleeve. If the tolerance on the depth of the hole is relatively small then these parts must be made to a high degree of accuracy if the work tolerance is to be maintained.

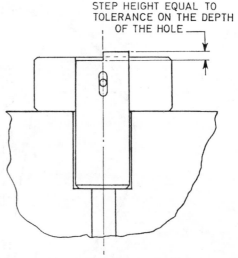

Fig. 5.13.   Hole depth gauge.

5.7 GAUGING OF LARGE DIAMETERS

The gauges used for determining the diameter of large workpieces are not, strictly speaking, limit gauges although they may be used as such.

5.71 Gauging Large Bores

The gauge used for this purpose is a 'pin' gauge whose length has been accurately determined using a length bar measuring machine (see Chapter 3).

The length of the pin gauge should be smaller than the diameter of the bore to be gauged. It is inserted in the bore, and the amount of rock about point A measured with a flexible steel rule or tape; see Fig. 5.14.

From Fig. 5.14 it is seen that $D = L + \delta$, and also if the angle of rock is small:

$$BC = \frac{W}{2}$$

As the angle in a semicircle is a right angle it can be seen that

$$L^2 + \left(\frac{W}{2}\right)^2 = (L + \delta)^2$$

Or

$$L^2 + \frac{W^2}{4} = L^2 + 2L\delta + \delta^2$$

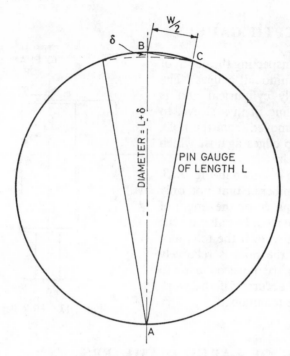

Fig. 5.14. Measurement of large bores.

The L^2 terms cancel and if the amount of rock is small δ is small, and $δ^2$ can be ignored.

$$\therefore \quad \frac{W^2}{4} = 2L\delta$$

$$\delta = \frac{W^2}{8L}$$

where L is the length of the pin gauge and W is the amount of rock

and as $D = L + \delta$

then $D = L + \dfrac{W^2}{8L}$

or $W = \sqrt{(D-L)8L}$

If the limiting values of D, and the known length L are inserted into this expression, then the limits allowed on the amount of rock may be derived from this final expression.

The accuracy which can be achieved by this method is of a much higher degree than can be obtained using an instrument such as a vernier calliper to make the measurement direct. Consider a gauge of 400 mm length exhibiting 50 mm rock.

Then

$$D = 400 + \frac{2500}{8 \times 400} \text{ mm}$$

$$= 400 + 0.78 \text{ mm}$$

The measurement of W is normally accurate to ± 1 mm and if a 1 mm error is assumed in W, then

$$D = 400 + \frac{(51)^2}{8 \times 400} \text{ mm}$$

$$= 400 + \frac{2601}{8 \times 400} = 400 + 0.81 \text{ mm}$$

Thus an induced error of 1 mm in the amount of rock produces an error of ± 0.03 mm in the diameter measurement, and on a diameter of 400 mm this would be difficult to improve upon with simple direct measuring equipment.

5.72 Measuring of Large External Radii

To carry out these measurements an angle piece may be made up, the included angle between whose faces is accurately known. This may be fitted with a micro-meter as shown in Fig. 5.15 or used in conjunction with a stepped setting block,

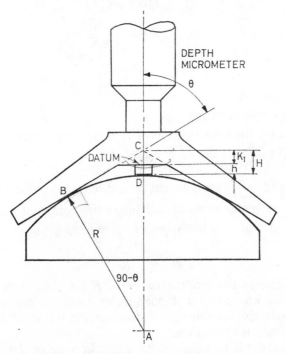

Fig. 5.15. Measurement of large external radii.

the height of the step being related to the limits of the radius under test. It is of interest that this type of equipment determines the radius of the workpiece, not its diameter, and can thus be used to determine the radius of a segment of a circle as in Fig. 5.15.

If θ is the semi-angle between the faces of the angle gauge, then, from Fig. 5.15,

$$\cos (90 - \theta) = \frac{AB}{AC} \text{ in which } AB = R$$
$$AC = R + H$$

$$\therefore \cos (90 - \theta) = \frac{R}{R + H}$$

$$(R + H) \cos (90 - \theta) = R$$

$$H \cos (90 - \theta) = R - R \cos (90 - \theta)$$

$$= R[1 - \cos (90 - \theta)]$$

$$\therefore R = \frac{H \cos (90 - \theta)}{[1 - \cos (90 - \theta)]}$$

In this expression for R the term $\cos (90 - \theta)$ is a constant for a given gauge.

$$R = \frac{HK}{(1 - K)}$$

and referring to Fig. 5.15 it is seen that

$$H = h + K_1$$

$$R = \frac{(h + K_1)K}{1 - K}$$

$$= (h + K_1) \frac{K}{1 - K}$$

in which $K = \cos (90 - \theta)$

θ = semi-angle of gauge

K_1 = distance from intersection of gauging faces to datum

H = height from radius to datum face

The values of K_1 and $\frac{K}{1 - K}$ are constant for a given gauge, and it is useful to have them stamped on the gauge as numerical values to avoid confusion.

If the expression for R is transposed to give h we get

$$h = \frac{R(1 - K)}{K} - K_1$$

and if in this expression the limiting values of R are used then the values of h obtained are the thicknesses of a stepped feeler gauge to use with the gauging unit, or alternatively the micrometer readings at the limits of R, assuming the micrometer reads zero at the datum.

Another alternative is to mount a dial gauge in place of the micrometer, and fit limit fingers to the dial gauge at the limiting readings of h.

It should also be noted that since this method reveals the error in the radius of the workpiece, it is doubled to give the error in the diameter of the work. It is thus a sensitive method of detecting errors in relatively large diameter workpieces.

5.8 MATERIALS FOR GAUGES

If a material is to be used successfully for gauge manufacture, it must fulfil certain requirements, either by virtue of its own properties, or by having these properties conferred upon it by manufacturing or a heat treatment process.

These requirements are:

(a) *Hardness*. To resist wear.

(b) *Stability*. Its size and shape must not change over a period of time.

(c) *Corrosion resistance*.

(d) *Machineability*. It must be easily machined into the required shape and to the required degree of accuracy and surface finish.

(e) *Low coefficient of linear expansion*. A limit gauge is often subject to a considerable amount of handling compared with the workpiece. For this reason it is desirable to have a low expansion coefficient but it should be noted that parts of the gauge which are to be held in the hand should have low thermal conductivity. It is recommended, for example, that plug gauges consist of steel gauging units held by tapers in ebonite or other plastic handles.

It is perhaps fortunate that a suitable material for gauge manufacture is a relatively inexpensive good quality high carbon steel. Suitable heat treatment can produce a high degree of hardness coupled with stability, and at the same time it can be readily machined and brought to a high degree of surface finish.

5.81 Heat Treatment of Limit Gauges

A high carbon steel is fully hardened by heating to 730° and quenching in water. This will give a hardness of approximately 64 on the Rockwell 'C' scale, but it will also make the steel extremely brittle. It is necessary to temper the gauge to reduce the brittleness, but not to make it so soft as to reduce its wear resistance. At the same time the tempering treatment can be used to stabilize the material and relieve any internal stresses which may distort it over a period of time.

A tempering temperature of 200°C will reduce the brittleness so that the gauge is not likely to chip and the hardness value will be Rockwell 'C' 58. If this temperature of 200°C is maintained over a period of 8 to 10 hours it will also make the gauge extremely stable.

Screw thread gauges are particularly fragile and prone to damage if roughly handled, and these gauges should be 'let down' further at a temperature of 240°C to give a Rockwell hardness of 'C' 52.

5.82 Other Materials for Limit Gauges

If a large piece of high carbon steel is water-quenched it will probably crack. If it is oil-quenched it will not attain a high enough degree of hardness to resist wear. Large gauges should therefore be made of an oil-hardening tool steel, usually of a nickel–chrome alloy.

Where expansion due to temperature is of particular importance as in the use of long precision gauges a material called *Invar*, an alloy containing 36% nickel, may be used. Invar has an expansion coefficient of less than 1×10^{-6} per °C but is unstable over a long period. It has been found that a 42% nickel alloy, known as *Elinvar*, is much more stable and has an expansion coefficient of 8×10^{-6} per °C which is still remarkably small.

Frequently a low carbon steel is used for gauge manufacture and is hardened by case-hardening. However, the hardening process is relatively expensive as it consists of carburizing, core-refining, case-hardening, and tempering. Such a gauge will probably be wholly satisfactory and the choice between high carbon steel, hardened right through, and a case-hardened low carbon steel is largely dictated by personal taste, material availability and economics.

Direct electro-plating of chromium on to steel has been developed, not for decorative purposes, but because chromium is extremely hard and wear resistant. This process is useful for reclaiming worn gauges, which are machined slightly undersize to correct any form errors which may have developed. They are then plated slightly oversize and ground or lapped to size.

During the war when good quality steels were required for other purposes, a number of glass gauges were used, particularly in the United States of America. It was claimed that they did not burr if dropped, but gave a clean chip and could still be used, or broke altogether. They could be stabilized and given a high surface finish. A peculiar advantage claimed was that they did not obscure the work and that visual inspection could be carried out in some cases while gauging was in progress. The Corning Glass Works, of Corning, N.Y. have reported a much longer life for glass compared with steel gauges, when used for gauging gun barrels.

5.9 SURFACE FINISH FOR GAUGES

Much can be done to reduce the initial wear rate of a gauge if its surface finish is good. A poor finish with a small number of high peaks is prone to more rapid wear than a finish having a large number of very small peaks, giving a large contact area. Thus a gauge should be finished by high quality grinding or lapping to give a C.L.A. value of not more than 0·10 μm.

CHAPTER 6

Machine Tool Metrology

6.1 THE NEED FOR MACHINE TOOL METROLOGY

In previous chapters we have been concerned with the foundations of metrology, and with the measurement of gauges. It will be clear that the term 'inaccuracy' signifies not only the suitability or otherwise of the dimensions of a gauge, but that it must also be interpreted to embrace such characteristics as quality of surface finish, and geometry. That a gauge may combine all these characteristics in suitable degree, is due to the skill of the gaugemaker and to the inherent quality and accuracy of the machine tools at his disposal.

A similar set of conditions applies to the production of components, except that a lower order of skill may be required and a higher rate of production achieved. To an increasing degree components are required to be of such accuracy that they may be assembled on a non-selective basis, the finished assemblies conforming to very stringent functional requirements. That this may be done economically is due in large part to the accuracy of the machine tools used in the production of the component parts.

The continuously increasing demands for highly accurately machined components has led to considerable research in machine tool design, and particularly towards means by which the geometric accuracy of machines may be improved and maintained. There has, therefore grown up a distinct field of metrology concerned primarily with the geometric tests of the alignment accuracy of machine tools under static conditions. An extension of this field, and a developing one at the present time, is the determination of the alignment accuracy of machine tools under dynamic loading conditions. This is a logical development, since both the designer and the user of machines are ultimately concerned with the behaviour and characteristics of a machine when under normal operating conditions.

This chapter will be confined to the static testing of alignments, in which well-established measuring methods are used. Dynamic tests are at the present time the subject of much research, and the reader is advised to refer to research papers on the subject for further information.

6.2 ALIGNMENT TESTS

The tests applied to machine tools, regardless of type, fall into well defined groups which may be summarized as follows:

(a) The level of installation of the machine in the horizontal and vertical planes.

(b) The main spindle is tested for axial movement and in the running truth of its axis.

(c) The bed-ways are tested for straightness and parallelism.

(d) The line of movement of members such as saddles and tables along bed-ways is tested.

(e) Practical tests in the form of the machining of test specimens, followed by careful and surface finish of the specimen.

The tests carried out under (e) above are in the nature of dynamic tests, in so far as the results of the measurement of the specimen reveal the behaviour of the machine under normal conditions of operation. That is, it is not sufficient that the machine should be satisfactory under conditions of static loading only, but that account should also be taken of the vibration and deflection of machine members under dynamic loads.

Much pioneering work in devising and evaluating suitable tests was carried out by Dr. G. Schlesinger, whose work *Testing Machine Tools* is a standard treatise on the subject.

At a subsequent date, a series of test charts were prepared jointly by the Institution of Mechanical Engineers and the Institution of Production Engineers, and it is these which are used, or form the basis of the standard tests applied by most machine tool makers, and constitute the procedure for the final inspection of the machine.

The accuracies specified in 'Acceptance Test Charts for Machine Tools' are recognized as being satisfactory for the types of machines to which they refer, but at the present time improved methods of making the measurements are available, and these will be included in this chapter.

6.21 Tests for Level of Installation

It cannot be too strongly emphasized that it is essential that a machine tool be installed in truly horizontal and vertical planes, and that this accuracy must be maintained. If, for example, we consider the case of a long bed lathe which is not installed truly horizontal, it is clear that the bed will undergo a deflection, either to produce a simple bend, or if the deflection is in two directions, a twist will be introduced.

It would thus follow that the movement of the saddle could not be in a straight line, and it would therefore be impossible to turn a true geometric cylinder on the lathe. Similar effects are obtained on other types of machines when the accuracy of installation is not of the required order.

The maintenance of the initial installation accuracy is dependent upon the type and thickness of foundation on which the machine is set. Many machines require no more than the normal thickness of workshop floor, say 150–200 mm of concrete laid on hardcore. In special cases, such as jig boring machines and high precision grinders, it is necessary to prepare a special concrete foundation of considerable depth, which additionally may be insulated from the surrounding floor.

The method of testing for level is shown in Fig. 6.1 in which is outlined the bed and bed-ways of a centre lathe.

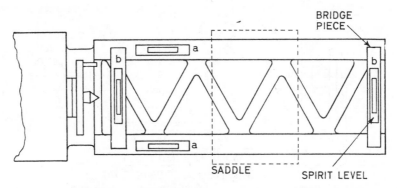

Fig. 6.1.   Testing a lathe bed for level and wind.

With the saddle at the approximate mid-span of the bed support feet, a precision level is placed at a–a to give the level in the longitudinal direction. The transverse test at b–b will usually require the use of a bridge piece to span the front and rear guide-ways. Preferably, readings should be taken simultaneously in each direction so that the effect of adjustments in one direction may also be observed on the level in the other. It will be noted that readings taken transversely will reveal any twist or 'wind' in the bed. That is, if the bed is out of level, readings of the bubble position should all be either plus or minus.

The process of correcting the error in level may be done by wedges and shims set at suitable points under the support feet or pads of the machine until an accuracy of 0·02 mm/metre is obtained.

The type of level suitable for this work has a senstivity of 0·04 mm/metre and has a vee base, so that direct contact may be made with the inverted vee ways of the machine bed.

The bridge piece to span the bed for the transverse test may have a flat base for use on flat-bed lathes, but a vee bridge is generally considered better. The manufacture of such a bridge is somewhat simplified if hardened and ground cylinders form the vees.

6.22 Spindle Tests

Considering again the testing of a centre lathe, the running truth of the internal taper at the front end of the spindle is of vital importance for the production of accurate work. The equipment required for this test consists of:

(*a*) Hardened ground parallel test bar having a concentric taper shank which is a close fit to the spindle nose taper.

(*b*) A dial test indicator calibrated in 0·01 mm.

The method is shown in Fig. 6.2.

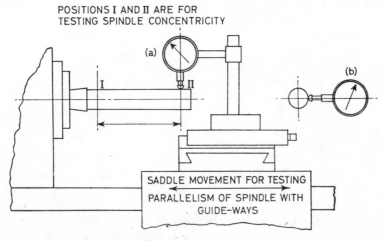

Fig. 6.2. Tests for spindle concentricity and alignment with guide-ways.

Two readings, at I and II, are taken of the running truth of the test bar, which obviously must itself be a highly accurate one in all respects. Over a distance such as 300 mm between I and II, a difference of running truth is allowed At I the error should not exceed 0·01 mm and at II, 0·03 mm.

A most important test is that of the parallelism of the spindle axis with respect to the bed-ways, in both the vertical and horizontal planes. The method of test is shown in Fig. 6.2. Any error in parallelism which is revealed when the saddle is moved along the bed should not exceed 0·02 mm/300 mm in either plane. In the vertical plane, shown at (*a*), the reading may be in the plus direction at the free end of the test bar. In the case of the horizontal plane, shown at (*b*), any error should. incline the test bar towards the direction of the tool pressure. By permitting only these directions of error, the tendency is for the tool pressures to correct them.

The axial slip or float of the spindle is tested as in Fig. 6.3, the dial test indicator being firmly mounted at any suitable position on the machine.

The test plug must be a snug fit into the spindle nose taper, and have its outer end face ground flat and square to the taper axis. The line of movement of the measuring plunger is then arranged to lie on the spindle axis of rotation. The slip should not exceed 0·01 mm.

It is important to distinguish between the necessary end play in the spindle, due to the clearance between bearing surfaces, and axial slip. The latter is due to errors in squareness of the abutment faces or locating faces of end bearings with respect to the axis of rotation, and produces a cyclic end-wise movement of the spindle.

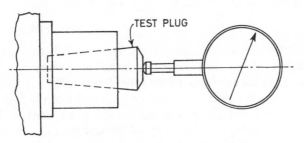

Fig. 6.3. Test for axial slip of lathe spindle.

6.23 Tests for Straightness and Flatness

It will be appreciated that for a carriage to move along a straight line in both vertical and horizontal planes, the controlling guide-ways must themselves be straight. Tests for this condition may be carried out in several ways, the most convenient of which are by precision level and by the auto-collimator. It is the latter method which will be discussed here, but the method of tabulating and using the results of individual measurements is similar in each method.

The principle of measurement by the auto-collimator has been dealt with in Chapter 4, but the method of determination of straightness and flatness is dealt with now.

Assume that the straightness of a lathe bed 2 m in length is to be measured. The general arrangement of measurement would be as in Fig. 6.4, the auto-collimator being set up independently of the lathe bed, about ½ m from one end, the parallel beam from the instrument being projected along the length of the bed. A particularly rigid support, preferably of the tripod type, is required for this. Assuming the bed to have flat-ways, the plane reflector is set on to the end of the bed nearer the instrument and a reflection obtained from it such that the image of the cross-wires of the collimator appear nearer the centre of the field. The reflector is then moved to the other end of the bed, and provided the general line of movement of the reflector has been reasonably parallel to the optical axis of the instrument, then the image of the cross-wires will appear in the field of the eyepiece at this position of the reflector also. This procedure ensures that reflections at intermediate positions will be within the field, and is thus an approximate check on the level of the bed in the horizontal plane.

119

A straight-edge should now be set down on the bed, to ensure that the reflector is stepped along it in a straight line.

Assume that the distance between the support feet of the reflector is 103·5 mm, and that the interval length at which measurements are taken is also 103·5 mm.

Now, since 1 min of arc

$$= \frac{2\pi}{360 \times 60} \text{ radians}$$

then, on a base length of 103·5 mm, 1 min of arc $= \frac{2\pi}{360 \times 60} \times 103 \cdot 5$ mm

$$= 0 \cdot 03 \text{ mm}$$

That is, each tilt of 1 min of arc of the reflector as it is stepped along the bed-way corresponds very closely to a rise or fall of the guide-way surface of 0·03 mm.

Having ensured that an image of the cross-wires will be received by the auto-collimator when the reflector is set at the end positions of the bed, the reflector is now set at the forward end of the bed, nearest the instrument, to begin the series of readings. This condition, and those for subsequent readings, is shown in Fig. 6.4 in which the rise and fall of the bed surface is greatly exaggerated.

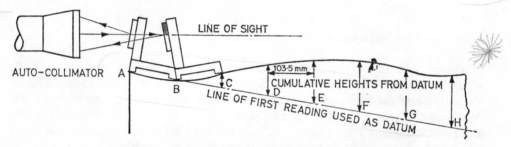

Fig. 6.4. Auto-collimator used for checking the straightness of lathe-bed guide-ways.

With the reflector set at A–B, the setting wires in the auto-collimator eyepiece are moved to straddle symmetrically the image of the horizontal cross-wire, by the suitable rotation of the micrometer drum, and the micrometer reading is noted. The reflector is then moved 103·5 mm to the position B–C and a second reading is taken on the micrometer drum. Successive readings at C–D, D–E, E–F, etc., are taken until the length of the bed has been stepped along. A second set of readings should now be obtained by stepping the reflector in the reverse direction along the bed, to reveal any serious errors in the first set of readings. Assuming none have occurred, the mean values of each set of readings may now be recorded, and these represent the angular positions of the reflector, in seconds relative to the optical axis of the auto-collimator at each of its positions along the bed.

The method of tabulation of the results of measurement are shown on p. 121.

Column 1 gives the position of the plane reflector at 103·5 mm intervals along the bed. Column 2 gives the mean reading of the auto-collimator to the nearest second. In practice it is possible to observe sub-divisions of seconds, and this should be done. Column 3 gives the differences of each reading from the first. In column 4 these differences are converted to the corresponding linear rise or fall, on the basis of 1 sec of arc = 0·0005 mm per 103·5 mm. The second zero introduced at the head of column 4, when associated with the previous zero in this column, represents the heights of the two feet of the reflector support mounting when in its original position. Column 5 gives the heights of the support feet of the reflector above the datum line drawn through their first position. That is, the values in column 5 are obtained by successively adding, algebraically, the values in column 4. This is necessary because the individual heights obtained in column 4 are the heights of the back feet of the support relative to the front feet in a given position and not relative to the datum.

1	2		3	4	5	6	7
				Rise or		*Adjustment*	*Errors*
			Difference	*Fall in*	*Cumulative*	*to bring*	*from*
Position	*Reading*		*from 1st*	*Interval*	*Rise or*	*both ends*	*Straight*
on			*Reading*	*Length*	*Fall*	*to Zero*	*Line*
Surface	*min*	*sec*	*(sec)*	*(0·001 mm)*	*(0·001 mm)*	*(0·001 mm)*	*(0·001 mm)*
0	2	10	0	0	0	0	0
1	2	10	0	0	0	− 2·0	−2
2	2	12	+2	+1·0	+ 1·0	− 4·0	−3
3	2	15	+5	+2·5	+ 3·5	− 6·0	−2·5
4	2	17	+7	+3·5	+ 7·0	− 8·0	−1·0
5	2	18	+8	+4·0	+11·0	− 10·0	+1·0
6	2	17	+7	+3·5	+14·5	− 12·0	+2·5
7	2	15	+5	+2·5	+17·0	− 14·0	+3·0
8	2	13	+3	+1·5	+18·5	− 16·0	+2·5
9	2	9	−1	−0·5	+18·0	− 18·0	0
10	2	12	+2	+1·0	+19·0	− 20·0	−1·0
11	2	14	+4	+2·0	+21·0	− 22·0	−1·0
12	2	16	+6	+3·0	+24·0	− 24·0	0

The total rise in the surface of the bed over a $1\frac{1}{4}$ m length from a datum along the line of the first reading is 24 μm. In column 6 this total rise is proportioned over the twelve readings taken, i.e. in increments of 24/12 = 2 μm. These values (column 6) are subtracted from the values in column 5 to give the errors (column 7) in the bed from a straight line joining the end points and within which the series of readings were obtained (i.e. it is as though a straight-edge were laid along the bed profile and touching the end points of the test surface when they are in a horizontal plane). The rise and fall of the surface relative to the straight-edge would be the values given in column 7.

A graphical representation of this is shown in Fig. 6.5 in which the values

121

given in columns 5, 6, and 7 are plotted. In the graph of cumulative errors a straight line has been passed through the end points, and represents the straight line connecting the ends of the bed. In the graph of straightness errors, this line has been used as the axis, and thus the values plotted in the previous graph have the same relationship to it.

It is important to note that the increasing values for the readings given in column 2 of the table indicate the increasing angle of tilt of the top of the reflector towards the optical axis of the auto-collimator. Increasing readings have therefore

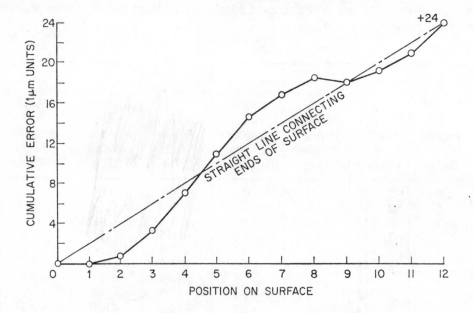

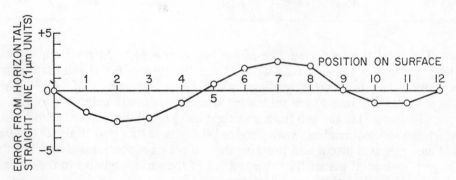

Fig. 6.5. Graphs of cumulative error and actual error in a machine bed, determined using an auto-collimator.

122

indicated positive (+) values for the linear rise and fall, and vice versa. The lathe bed is thus both concave and convex along its length relative to the datum line joining its end points.

The test described relates to a flat-bed lathe, but the method applies also to a bed with vee guide-ways. In this case, however, the plane reflector mount must be supported on a carriage having a vee accurately ground in its base, to suit the vee of the bed. The apex of the carriage vee should be relieved so that contact is made only at the sides of the vee. With this arrangement, the straightness of the bed in the horizontal plane can be determined, as well as in the vertical (i.e. the tilt about the vertical axis indicates changes in the angle of the reflector in the horizontal plane).

As in the previous test, the reflector is stepped along the bed in interval lengths of 6 in but in this case the auto-collimator tube is rotated 90° in its housing, to a pre-set stop, so that the pair of setting wires in the eyepiece are vertical. Changes of position of the image of the vertical member of the cross-wires are then read on the micrometer drum, and would be recorded in column 2.

Associated with the error in the straightness of a lathe bed may be any cross-wind which exists due to each way of the bed having different errors in straightness, or, if straight, lying in non-parallel planes. This condition cannot be detected with an auto-collimator, since the reflector would be merely rotating in its own plane as it was stepped along the bed. The most practical and convenient method of test is to step a precision level, laid transversely across the bed, along its length. If necessary, a bridge piece should be used as the level carriage, both to span the width of the bed, and to accommodate the vee guideways.

6.3 TESTS FOR SQUARENESS

In many machine tools, for example, drilling machines and vertical milling machines, an essential accuracy lies in the squareness of the spindle axis with the plane of the table. Fig. 6.6 illustrates the principle, in which a test bar supports a dial test indicator at a suitable radius, say 150 mm.

With the measuring plunger bearing directly but lightly on the machine table, the spindle is rotated slowly by hand through 360°, dial indicator readings being taken at 180° in planes perpendicular to each other. In the case of drilling machines, the permitted error is of the order of 0·05 mm/300 mm, and in direction such that the front edge of the table is inclined upwards. That is, the force due to the downward traverse of the drill tends to correct the error.

In the case of vertical milling machines, the full test would apply only to fixed cutting head machines. A test in a direction perpendicular to the longitudinal table axis only is applicable to swivelling head machines, and an error of 0·02 mm/300 mm is applicable, the front edge of the table inclining upwards.

A test of this nature is known as the 'turn round method', and proves to be a very sensitive one.

In some classes of machine tools, the accuracy of construction called for requires optical methods for its testing. This applies especially to jig-boring machines, but is also applicable to milling machines of both vertical and horizontal types.

Fig. 6.7 illustrates the conditions of test for the squareness of the transverse table ways with the face of the column.

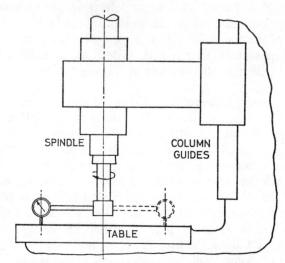

SPINDLE

COLUMN GUIDES

TABLE

Fig. 6.6.   Testing a drilling-machine spindle for squareness with the table.

When using an optical method for such a test, and bearing in mind that the axis of the incident beam from the auto-collimator forms the measuring datum, it is clear that this must be turned accurately through 90°. It is done by means of an optical square. This is a special case of a prism, such that, regardless of the angle at which the incident beam strikes the face of the prism, then by internal reflection the beam is turned through 90°.

Assume, then, that the transverse ways and the column face are perfectly

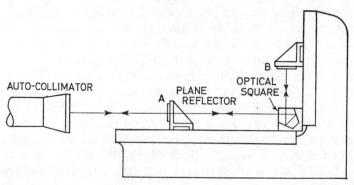

AUTO-COLLIMATOR

B

OPTICAL SQUARE

A   PLANE REFLECTOR

Fig. 6.7.   Auto-collimator and optical square being used to test the squareness of machine slides.

124

straight. It is necessary only to take two readings; one with the reflector at position A, and a second at position B, the optical square being set down at the intersection of the two surfaces when the reading at B is taken. The difference between the two readings would then be the squareness error, note being made of its direction.

The test becomes more complex, however, if the straightness errors in the two surfaces are considered. In this case, it is necessary to carry out a straightness test on each surface, as shown in section 6.23. The angle formed by the mean straight lines passing through each surface profile will then indicate the mean angle between the surfaces.

A squareness error, which reveals itself as an axial movement, is that of the thrust face and collars of a lead screw. Unless these are perfectly square to the axis of lead-screw rotation, a cyclic end-wise movement is set up which is of the same nature, and due to the same reasons as the axial slip in a main spindle. The condition is shown in Fig. 6.8.

The two methods of testing are:

(*a*) by dial test indicator;

(*b*) by auto-collimator.

Method (*a*) is similar in all respects to that used for the axial slip of main spindles and the error which would be tolerated in the case of lathe lead screws would be of the order of 0·01 mm.

The importance of this test may be appreciated when it is realized that axial slip will cause a thread having a periodic pitch error to be cut on the lathe.

Method (*b*) provides an interesting application of the auto-collimator. The axial oscillations of the lead screw are converted to angular movements of a plane reflector ball located into the centre hole in the end of the lead screw, and mounted on a cross-strip hinge (Fig. 6.8). The angular movement of the reflector during one revolution of the lead screw is observed in the auto-collimator eyepiece.

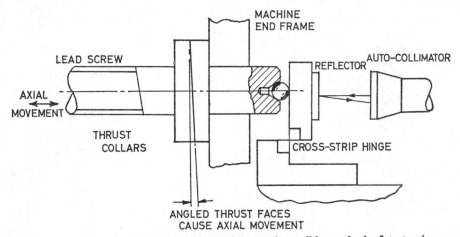

Fig. 6.8. Lead screw with angled thrust faces and possible method of test using an auto-collimator.

6.4 PRACTICAL TESTS

As previously noted, these are of a type designed to reveal the combined effects of several possible errors both in the alignment accuracy of the machine, and in its rigidity. They involve the machining of a test piece under prescribed conditions of cutting speed, feed rate, depth of cut, and tool geometry. The test piece is then measured for its geometry and surface finish, and the results compared with standards for these features.

CHAPTER 7

Gear Measurement

7.1 INTRODUCTION

As technology has progressed from the Industrial Revolution to the present day, the need for closer control over the accuracy of systems used for transmitting the power made available has also progressed. Probably the most used means of transmitting power and multiplying torque is through the medium of gear trains. It is obvious that the strength of gear teeth has had to improve to meet increased loads, but this is a design problem which is not a primary concern of this book. However, it is also a requirement of a gear train that it shall have a constant velocity ratio. Variations in velocity ratio can cause a cyclic fluctuation of tooth loading which gives rise to (a) fatigue, leading to tooth failure; and (b) noise.

The noise problem is of interest if one considers the development of the automobile. Early automobiles had rudimentary exhaust silencers and the resulting engine noise caused most of the other mechanical noises to be overlooked. Efficient exhaust silencing made mechanical noises from the gear-box more apparent. This was silenced by the use of helical gears and closer control in their manufacture. The gear noise was reduced and carburettor intake noise became significant which, when reduced by efficient air cleaners and intake silencers, enabled rear axle 'whine' to make its presence felt. The use of spiral bevels and hypoid gears, again with closer manufacturing controls, reduced this and the valve timing gears again required attention. By this time, exhaust and intake silencers were improved and the whole cycle started again.

Thus a major item of development in the motor vehicle has been the development of efficient gears, and this only considers one commodity. If one considers this work applied to all of the mechanisms which rely on geared systems to transmit power, the importance of the subject of gear measurement becomes immediately apparent.

7.2 SCOPE

A few of the different types of gears required by modern industry have been mentioned above. Within the confines of this work it is proposed to deal only with involute gears of straight tooth (spur) and helical types. These constitute a large proportion of the gears in use today, bevel gears, spiral bevels, and hypoid gears being topics for works of a more specialist character. Cycloidal gears are used but little in modern engineering. Their main use is in horological work, which again the authors consider is outside the scope of this work.

The choice of the involute for the flank curve of gear teeth has two great advantages for general engineering.

(*a*) The velocity ratio of a pair of involute gears is constant, regardless of errors or variations in centre distance.

(*b*) An involute rack has straight teeth. This enables the complex involute form to be generated from a relatively simple cutter.

It is therefore necessary to consider the involute curve in some detail.

7.3 THE INVOLUTE CURVE

An involute is the locus of a point on a straight line which rolls around a circle without slipping. An alternative definition is: the locus of a point on a piece of string which is unwound from a stationary cylinder.

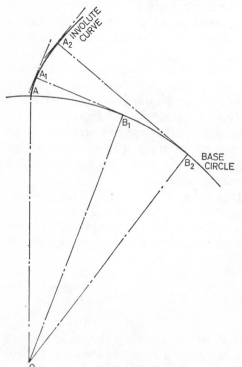

The curve is therefore as shown in Fig. 7.1.

From the figure it is seen that the length of the generator is equal to the arc length of the base circle from the point of tangency to the origin of the involute at A.

i.e. $A_1B_1 = $ arc AB_1

$A_2B_2 = $ arc AB_2 and so on.

Further, the tangent to the involute at any point, e.g. A_2, is perpendicular to the generator at that point.

Notice also that the shape of the involute depends entirely on the diameter of the base circle from which it is generated. As the base circle increases, so the curvature of the involute decreases, until the limit is reached for a base circle of infinite diameter, i.e. a straight line, when the involute is a straight line.

Fig. 7.1.   The involute curve.

7.4 THE INVOLUTE FUNCTION

The involute function of an angle may be defined as the angle made by the radius to the origin of the involute and the radius to the intercept of the generator with the involute. This is the involute function of the angle between the radius to the point of tangency of the generator and the radius to the intercept of the generator and the involute.

This apparently complex statement is better described graphically in Fig. 7.2.

In Fig. 7.2:

AOC is the involute function of COB.

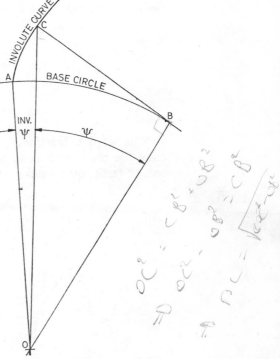

Fig. 7.2.   The involute function of an angle.

From the diagram (7.2):

$$BC = \sqrt{OC^2 - OB^2}$$

$$\tan \psi = \frac{\sqrt{OC^2 - OB^2}}{OB}$$

But from Fig. 7.1:

$$\text{arc } AB = BC$$

$$\therefore \frac{AB}{OB} = \psi \text{ radians} + \text{inv } \psi \text{ radians} = \frac{BC}{OB}$$

$$\therefore (\psi + \text{inv } \psi) \text{ radians} = \tan \psi$$

$$\text{inv } \psi = (\tan \psi - \psi) \text{ radians}$$

i.e. *the involute function of an angle is the difference between the tangent of the angle and the angle in radians.*

This term of involute geometry has been dealt with separately as it is of particular importance in the work to follow.

7.5 DEFINITIONS AND STANDARD PROPORTIONS

A single tooth of a gear is made up of portions of a pair of opposed involutes.

The teeth of a pair of gears in mesh contact each other along a *line of action* which is the common tangent to their base circles as shown in Fig. 7.3. As this is the common generator to both involutes, the load, or pressure between the gears is transmitted along this line. The angle between the line of action and the common tangent to the pitch circles is therefore known as the *pressure angle, ψ.*

From Fig. 7.3:

$$\frac{\mathrm{OB}}{\mathrm{OC}} = \cos \psi = \frac{R_b}{R_p}$$

$$\therefore R_b = R_p \cos \psi$$

or $D_b = D \cos \psi$ where $D_b =$ dia. of base circle

$D =$ dia. of pitch circle

$\psi =$ pressure angle

The standard values for pressure angle are $14\frac{1}{2}°$ and $20°$, of which $20°$ is becoming the most used as it gives stronger teeth and allows gears of smaller numbers of teeth to be made, without interference with mating teeth.

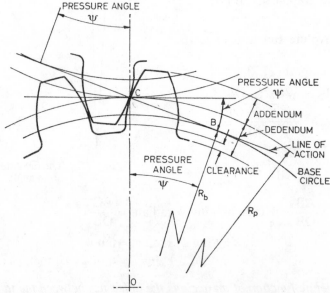

Fig. 7.3. Pair of spur gears in mesh, showing terms referred to in the text.

Diametral pitch P is the number of teeth per inch of pitch circle diameter. This is a hypothetical value which cannot be measured, but it is most important as it defines the proportions of all gear teeth.

$$P = \frac{N}{D}$$

Module M is the reciprocal of P, i.e.

$$M = \frac{D}{N}$$

This method of fixing tooth proportions is in common usage in countries using the metric system where M is made a whole number of millimetres.

130

Circular pitch CP is the arc distance measured around the pitch circle from the flank of one tooth to a similar flank in the next tooth.

$$\therefore CP = \frac{\pi D}{N} \text{ but } \frac{D}{N} = \frac{1}{P} = M$$

$$\therefore CP = \frac{\pi}{P} = \pi M$$

Base pitch P_b is the arc distance measured around the base circle from the origin of the involute on one tooth to the origin of a similar involute on the next tooth.

$$P_b = CP \cos \psi = \pi M \cos \psi$$

Addendum is the radial distance from the pitch circle to the tip of the tooth. The nominal value is:

$$\text{Addendum} = \frac{1}{P} = \text{Module}$$

This may be varied to avoid interference.

Clearance is the radial distance from the tip of a tooth to the bottom of a mating tooth space when the teeth are symmetrically engaged. Standard values are:

$$\text{Clearance} = \frac{0 \cdot 157}{P} \text{ or } \frac{0 \cdot 250}{P} \text{ or } \frac{0 \cdot 400}{P} = 0 \cdot 157\, M \text{ or } 0 \cdot 250\, M \text{ or } 0 \cdot 400\, M$$

The value used depends on the type of gears and their application.

$0 \cdot 157\, M$ is normally used for $14\frac{1}{2}°$ pressure angle gears to Browne and Sharpe standards.

$0 \cdot 250\, M$ is normally used for Class A_2, B, C, and D gears.

$0 \cdot 400\, M$ is normally used for Class A_1 precision ground gears.

Dedendum is the radial distance from the pitch circle to the bottom of the tooth space.

$$\text{Dedendum} = \text{Addendum} + \text{Clearance}$$

$$= \frac{1}{P} + \frac{0 \cdot 157}{P} = \frac{1 \cdot 157}{P} = 1 \cdot 157\, M$$

$$\text{or} = \frac{1}{P} + \frac{0 \cdot 250}{P} = \frac{1 \cdot 250}{P} = 1 \cdot 250\, M$$

$$\text{or} = \frac{1}{P} + \frac{0 \cdot 400}{P} = \frac{1 \cdot 400}{P} = 1 \cdot 400\, M$$

Blank diameter. The diameter of the blank is equal to the pitch circle diameter plus two addenda:

131

$$\text{Blank diameter} = D + 2M$$

$$\text{but } D = NM$$

$$\therefore \text{Blank diameter} = NM + 2M = (N + 2) \times \text{Module or } \frac{(N+2)}{P}$$

Tooth thickness is the arc distance measured along the pitch circle from its intercept with one flank to its intercept with the other flank of the same tooth.

Nominally, tooth thickness $= \frac{1}{2}CP$

$$= \frac{\pi}{2DP} \text{ or } \pi \times \frac{\text{Module}}{2}$$

In fact the thickness is usually reduced by an amount to allow for a certain amount of backlash and may be changed owing to addendum correction.

Backlash is the circumferential movement of one gear of a mating pair, the other gear being fixed, measured at the pitch circle, bearing clearances being eliminated.

It will be noted from the above definitions that a spur gear can be completely specified in terms of

      (a) number of teeth N;

      (b) diametral pitch P or module M;

      (c) pressure angle ψ.

In the work on gear measurement which follows the expressions derived will, where possible, all be reduced to functions of these dimensions.

7.6 HELICAL GEARS

A helical gear has involute teeth which are not cut parallel with the axis of rotation, as on spur gears, but at an angle known as the helix, or spiral, angle to it.

Helical gears are normally used to transmit power between parallel shafts. They provide a much smoother and quieter action than spur gears, owing to the fact that at any instant a number of teeth are engaged. Further, each tooth is engaged over a short length of its flank at a given time, and the engagement is taken up and released gradually.

Thus any measurements made can be in one of three planes:

(a) Normal to the tooth flank—denoted by subscript n.

(b) Normal to the axis of rotation—known as the transverse plane and denoted by subscript t.

(c) Parallel with the axis of rotation—known as the axial plane and denoted by subscript a.

In this work the measurements mainly considered will be those in the transverse and normal planes. Consider a rack of transverse pitch CP_t as in Fig. 7.5

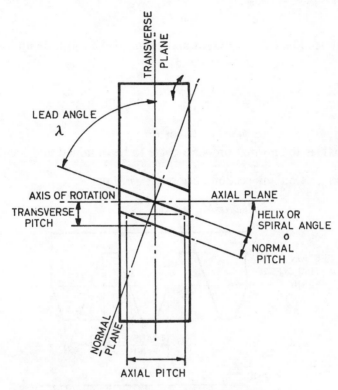

Fig. 7.4. Terms used in text dealing with helical gears.

Helix angle σ

Transverse pressure angle ψ_n

Transverse circular pitch $= CP_t = \pi M_t$

From the diagram (Fig. 7.5):

$$\frac{AB}{AC} = \cos \sigma$$

$$\therefore \ AB = AC \cos \sigma$$

$$CP_n = CP_t \cos \sigma$$

As circular pitch $= \pi \times$ Module

$$\text{then } \pi M_n = \pi M_t \cos \sigma$$

$$\text{and } M_n = M_t \cos \sigma$$

133

Now the angle of the flank face of the rack in any plane is the pressure angle of the system in that plane.

Considering again Fig. 7.5:

$$\tan \psi_n = \frac{DE}{DF}$$

For any section DF is constant and equal to the tooth depth but

$$DE = D_1E_1 \cos \sigma$$

$$\therefore \tan \psi_n = \frac{D_1E_1 \cos \sigma}{DF} \text{ and } \tan \psi_t = \frac{D_1E_1}{DF}$$

$$\therefore \tan \psi_n = \tan \psi_t \cos \sigma$$

This enables the normal pressure angle to be obtained and hence the normal base pitch P_{bn}, as on any section base pitch $= \pi M \cos \psi$.

$$\therefore P_{bn} = \pi M_n \cos \psi_n$$

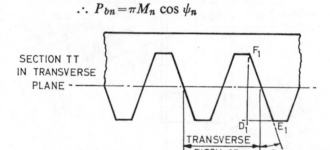

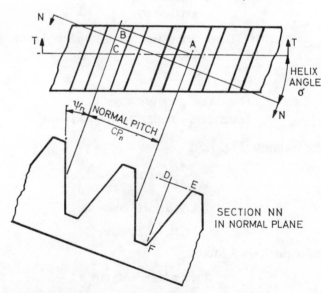

Fig. 7.5.   Normal and transverse sections of basic rack.

7.7 UNDERCUTTING IN GEAR TEETH AND ADDENDUM MODIFICATION

If in the design of a gear train a pinion is made too small, interference will occur between the mating pair. The tooth of the gear or rack will tend to 'dig in' to the root of the pinion, causing rough running or fracture. If a pinion is cut by a generating process in which interference would occur, the teeth become undercut and weakened. The condition occurs when the line of action extends beyond the point of tangency to the base circle.

Consider a pinion being cut by a rack as shown in Fig. 7.6, in which the tip of the rack cutter tooth extends beyond the point of tangency of the line of action by an amount S.

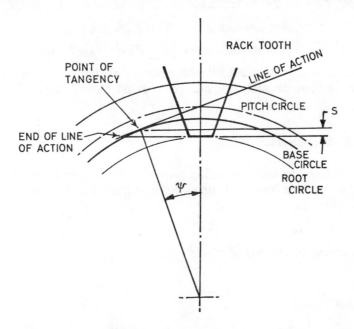

Fig. 7.6. Conditions giving rise to undercutting in gear manufacture.

This condition would give rise to undercutting and can be overcome by displacing the rack outwards a distance S. Thus, to compensate for this, the wheel addendum must be reduced by an amount K_w and the pinion addendum increased by an amount K_p to compensate.

The values K_p and K_w are called the addendum coefficients for the pinion and wheel and are directly related to S by $S = KM$, where $M = $ Module.

B.S. 436: 1940 gives two conditions for the calculation of K_p and K_w.

(a) If $(N+n) \sec^3 \sigma \geqslant 60$, then $K_p = 0.4 \left(1 - \dfrac{n}{N}\right)$

or $K_p = 0.02 \, (30 - n \sec^3 \sigma)$ whichever is the greater

and $K_w = -K_p$

(b) If $(N+n) \sec^3 \sigma < 60$, then $K_p = 0.02 \, (30 - n \sec^3 \sigma)$

and $K_w = 0.02 \, (30 - N \sec^3 \sigma)$

In the above expressions N and n are numbers of teeth in wheel and pinion respectively and σ is the helix angle for helical gears, thus for straight tooth spur gears $\sigma = 0$ and $\sec^3 \sigma = 1$.

Knowing $K_p = K_w$ the addenda for the gears may be found from:

Pinion addendum $= M_n(1 + K_p) = M_n + K_p M_n$

Wheel addendum $= M_n(1 + K_w) = M_n + K_w M_n$

Note that in each case the nominal addendum M is changed by an amount KM which is the amount the rack is displaced to avoid undercutting.

7.8 GEAR MEASUREMENT

The methods of testing and measuring gears depend largely on the class of gear, the method of manufacture, and the equipment available. Briefly, gear measurement can consist of:

(a) General tests.

(b) Measurement of individual elements.

7.81 General Tests

Included in this section are rolling tests, in which the gear is compared with a hardened and ground master gear, tooth thickness measurements, and measurements over rollers. The two latter may seem out of place in general tests, but when a gear is being cut, often complete reliance is placed on cutter and indexing accuracy, each of which influence form and pitch, and these are the only tests available. However, the authors feel that more attention should be paid to cutter accuracy and to the setting of the machine. For instance, in hobbing a gear of small module, the only tests available are optical projection and rolling tests. These frequently reveal errors in form which can be due only to either a hob of incorrect form, or to the hob being incorrectly set, i.e. not inclined accurately through the helix angle of the gear. The hob, having straight flanks, can be readily checked by optical projection and this leaves only inaccuracy of setting to cause the form errors.

In the manufacture of gears of higher class, e.g. precision-ground gears and master gears, it is necessary to determine the accuracy of individual elements. Apart from tooth thickness, these include (*a*) pitch of teeth, and (*b*) form of teeth.

7.9 ROLLING GEAR TESTS

A common form of gear testing machine for performing these tests is the Parkson Gear Tester, shown in Fig. 7.7. It consists essentially of a base on which is mounted a 'fixed' carriage whose position can be adjusted to enable a wide range of gear diameters to be accommodated, but which is locked in use, and a moving carriage which is spring loaded towards the fixed carriage. On the fixed carriage is an arbor made to suit the bore of the master gear, or a similar spindle to suit the gear under test being mounted in a parallel plane on the moving carriage.

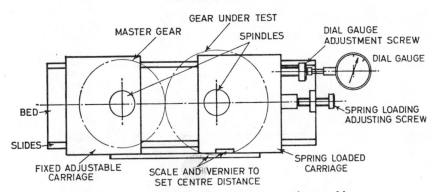

Fig. 7.7. Essentials of rolling-type gear testing machine.

If a pair of gears are spring loaded into close mesh and rotated, any errors in tooth form, pitch, or concentricity of pitch line, will cause a variation of centre distance. Thus, movements of the carriage, as indicated by the dial gauge, indicate errors in the production gear.

The method of operation is as follows:

(*a*) Using gauge blocks between the spindles set the dial gauge to read zero at the correct centre, distance and adjust the spring loading.

Note. Gauge blocks $= C - \dfrac{(D+d)}{2}$ where $C =$ centre distance and d and $D =$ diameters of spindles.

(*b*) Set limit marks on dial gauge.

(*c*) Mount the master gear and the gear to be tested, and note the variation in the dial gauge reading when the gears are rotated by hand. If it falls outside the limit marks, the gear is not acceptable.

This is the standard form of test used industrially under production conditions. However, the gear tester can be used to carry out more complex tests. For instance, by locking the moving slide at the running centre distance of the gears, and by fixing the master gear, the backlash can be determined by setting a dial gauge at the pitch line of the production gear. Also, at this setting, the gears can be checked for smooth running.

By arranging the machine to produce a trace, or graph of slide movements against gear rotation, the faults can be analysed. This can be accomplished by replacing or augmenting the dial gauge with a pick-up such as a 'Talymin' gauging head and feeding the output to the amplifying unit and recorder of a 'Talysurf' surface measuring instrument. Typical traces produced from such a set-up are shown in Fig. 7.8.

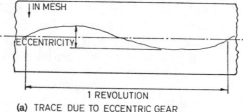

(a) TRACE DUE TO ECCENTRIC GEAR

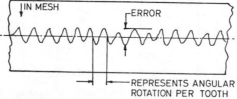

(b) INDIVIDUAL TOOTH ERRORS

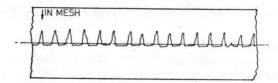

(c) <u>INTERFERENCE:</u> THE GEARS MESH NORMALLY UNTIL THE TIPS OF THE PINION MAKE CONTACT, WHEN THEY ARE FORCED OUT OF MESH

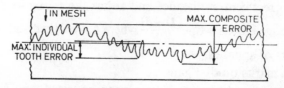

(d) COMPOSITE ERROR

Fig. 7.8. Typical traces produced by recording-type rolling test instrument.

It is usual for such errors to occur together and a compound trace of eccentricity and tooth form errors occurring on one gear would appear as shown in Fig. 7.8 (*d*).

It should be noted that when testing helical gears on the Parkson tester the vertical component of the normal force between the teeth has a tendency to cause one gear to lift, or ride up its arbor. For this reason, the master gear in these cases should be mounted between centres and rotated in such a direction that it absorbs the upward thrust, the equal and opposite downward component holding the tested gear down.

If no master gear is available, the instrument can still be used to check a pair of mating gears by running them together. It must be ensured that they do not have compensating errors. For example, equal eccentricities, if mounted in a particular angular relationship, would cancel out and show no error. If this method is used, the gears should be tested twice at relative angular positions of 180° to each other.

7.10 TOOTH THICKNESS MEASUREMENT—STRAIGHT TOOTH SPUR GEARS AND HELICAL GEARS

A frequently used instrument for measuring gear tooth thickness is the gear tooth vernier. It should be realized that, as the thickness varies from the tip to the base circle of the tooth, any instrument for measuring on a single tooth must

(*a*) measure the tooth thickness at a specified position on the tooth;

(*b*) fix that position at which the measurement is taken.

The gear tooth vernier therefore consists essentially of a vernier calliper for making the measurement W combined with a vernier depth gauge for setting the dimension h at which the measurement W is taken as in Fig. 7.9.

The positions at which the measurement can be made are normally limited to two.

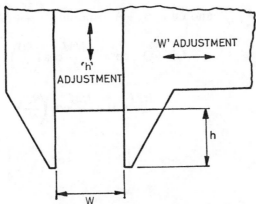

Fig. 7.9. Jaws of gear-tooth vernier.

7.101 Tooth Thickness at the Pitch Line

It should be noted that W is a chord AC, but the tooth thickness is specified as an arc distance ADC. Also h is the distance EB and this is slightly greater than the addendum ED.

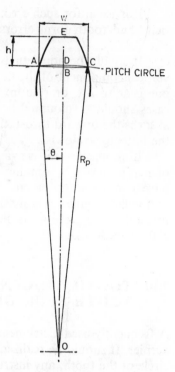

Referring to Fig. 7.10:

$$W = 2\,\text{AB}$$

In triangle ABO, $\text{AO} = R_p = \dfrac{D}{2}$

$$= \frac{NM}{2}$$

$$\theta = \frac{360°}{4N}$$

$$= \frac{90°}{N}$$

$$\sin\theta = \frac{\text{AB}}{\text{AO}}$$

$$\text{AB} = \text{AO}\,\sin\theta$$

$$= \frac{NM}{2}\sin\left(\frac{90}{N}\right)$$

Fig. 7.10. Measurement of chordal thickness at pitch line.

But $W = 2\,\text{AB}$

$$W = NM\sin\left(\frac{90}{N}\right) \qquad\qquad \dots (1)$$

Also from Fig. 7.10:

$$h = \text{OE} - \text{OB}$$

$$\text{and OE} = R_p + \text{Addendum} = \frac{NM}{2} + M$$

$$\text{OB} = \text{OA}\cos\theta = \frac{NM}{2}\cos\left(\frac{90}{N}\right)$$

$$\therefore h = \frac{NM}{2} + M - \frac{NM}{2}\cos\left(\frac{90}{N}\right)$$

$$h = \frac{NM}{2}\left[1 + \frac{2}{N} - \cos\left(\frac{90}{N}\right)\right] \qquad\qquad \dots (2)$$

and from (1) $W = NM\sin\left(\dfrac{90}{N}\right)$

In the above expressions it must be remembered that h and W are ideal values. Allowance must be made for h if the addendum is modified as in 7.10 and for W to accommodate the backlash required.

In the case of helical gears the tooth thickness in the transverse plane at the pitch line is not easy to measure but the normal tooth thickness can be determined. Within the limits of accuracy of this instrument, the normal tooth thickness W_n is the same as the tooth thickness of a corresponding virtual spur gear in which the number of teeth $N_v = N\sec^3\sigma$ where σ is the helix angle.

7.102   The Constant Chord

In the expressions for tooth thickness at the pitch line it is seen that the dimensions h and W are both dependent on the number of teeth. If a large number of gears for a set, each having different values of N, are to be tested, the separate calculations would become laborious.

Consider an involute tooth symmetrically in close mesh with a basic rack from as in Fig. 7.11. Regardless of the number of teeth, for a given size of tooth, i.e. value of M, contact would always occur at A and F. AF is known as the constant chord.

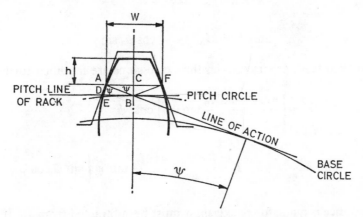

Fig. 7.11.   Measurement of tooth thickness at constant chord.

In Fig. 7.11:

$$BD = \tfrac{1}{4} \text{ circular pitch} = \frac{\pi M}{4}$$

In triangle ABD, $\dfrac{AB}{BD} = \cos \psi$

$$\therefore AB = BD \cos \psi = \frac{\pi M}{4} \cos \psi$$

In triangle ABC $\dfrac{AC}{AB} = \cos \psi$

$$\therefore\ AC = AB \cos \psi = \frac{\pi M}{4} \cos^2 \psi$$

$$\text{and}\ W = 2\ AC = \frac{\pi M}{2} \cos^2 \psi$$

Also, $\dfrac{BC}{AB} = \sin \psi$

$$\therefore\ BC = AB \sin \psi = \frac{\pi M}{4} \cos \psi \sin \psi$$

$$\text{and}\ h = M - \frac{\pi M}{4} \cos \psi \sin \psi \text{ for a value } W = \frac{\pi M}{2} \cos^2 \psi$$

The above expressions are for straight tooth spur gears. For helical gears, the normal measurement W_n is obtained by substituting values of normal module and normal pressure angle.

$$\therefore\ W_n = \frac{\pi M_n}{2} \cos^2 \psi$$

$$\text{and}\ h_n = M_n - \frac{\pi M_n}{4} \cos \psi_n \sin \psi_n$$

If corrected teeth are used, then the expressions are modified further to:

$$W_n = \left[\frac{\pi M_n}{2} + \frac{4k M_n}{2} \tan \psi_n \right] \cos^2 \psi$$

$$\text{and}\ h_n = (1 + k)M_n - \left[\frac{\pi M_n}{4} + \frac{4k M_n}{4} \tan \psi \right] \sin \psi \cos \psi$$

If allowance is made for backlash, it must be subtracted from W. If reduction in $W = \delta w$ and $b =$ backlash,

$$\delta w = b \sec \psi$$

An interesting point about the constant chord method is that it readily lends itself to a form of comparator which is more sensitive than the gear tooth vernier. Such an instrument is manufactured by W. E. Sykes, Ltd., and is illustrated in Fig. 7.12.

It consists of a pair of adjustable jaws having angles of $14\frac{1}{2}°$ or $20°$, depending on the application required, with a dial gauge probing down to the tip of the tooth, as shown. The instrument is set to a pair of master taper plug gauges which fix the position of the limit marks on the dial gauge.

142

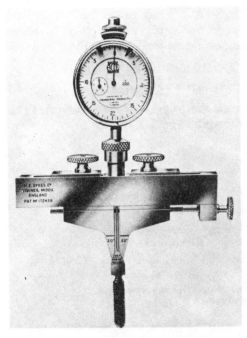

Fig. 7.12. Sykes Gear Tooth Comparator based on the constant chord method of measurement.

It should be noted that the fact that the jaws are at an angle provides an additional magnification of roughly 2 : 1, i.e. for a $14\frac{1}{2}°$ pressure angle, an error in tooth thickness of 0·01 mm shows up as 0·02 mm on the dial gauge.

7.103 The Base Tangent Method

Apart from the use of the Sykes Gear Tooth Comparator, it should be noted that both of the previous methods of measuring tooth thickness may be unsatisfactory using the gear tooth vernier, in that,

(*a*) the vernier itself is not reliable to closer than 0·05 mm or perhaps 0·025 mm with practice;

(*b*) the measurements depend on two vernier readings, each of which is a function of the other;

(*c*) measurement is made with an *edge* of the measuring jaw, not its face, which again does not lend itself to accurate measurement.

These problems can be overcome by measuring the span of a convenient number of teeth, as shown in Fig. 7.13.

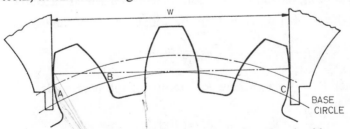

Fig. 7.13. Span measurement over a number of teeth with a vernier calliper.

It can be seen that this uses a single vernier calliper which overcomes the disadvantages mentioned above, except those inherent in the use of such an instrument, and this can be overcome by the use of more accurate equipment than a vernier calliper.

Consider a straight edge ABC of length AC, being rolled back and forth along a base circle as in Fig. 10.14. Its ends will sweep out opposed involutes $A_1.A.A_2.$ and $C_2.C.C_1.$ respectively. Measurements made across these opposed involutes by span gauging will be constant,

$$\text{i.e. } W = AC = A_1C_1 = A_2C_2 = \text{arc } A_0B_0$$

i.e. the arc length of the base circle between the origins of the involutes. (This condition should be compared with that shown at Fig. 7.1.)

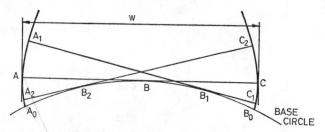

Fig. 7.14. Generation of a pair of opposed involutes by a common generator.

Therefore the position of the measuring faces is unimportant as long as they are parallel and on an opposed pair of true involutes. Applying this to gear teeth, it is in fact preferable to choose a number of teeth such that the measurement is made approximately at the pitch circle of the gear, for it is at this point that the tooth form is most likely to conform to a true involute.

Applying this principle to the conditions in Fig. 7.13, it is seen that

$$W = \text{arc } AB + \text{arc } BC$$

in which arc AB is the tooth thickness at the base circle and arc BC is ($S \times$ the base pitch), S being the number of tooth *spaces* over which measurement is made.

If N is the number of teeth, then the angular pitch of the teeth P_θ is given by

$$P_\theta = \frac{2\pi}{N} \text{ radians}$$

and base pitch $P_b = \dfrac{2\pi}{N} \times R_b$ where $R_b = $ radius of the base circle

Now $R_b = \dfrac{NM}{2} \cos \psi$ $\qquad\qquad M = \text{Module}$

$$\therefore P_b = \frac{2\pi}{N} \times \frac{NM}{2} \cos \psi$$

and if S tooth spaces are considered,

$$\text{arc } BC = \frac{2\pi}{N} \times S \times \frac{NM}{2} \cos \psi \qquad\qquad \dots (1)$$

To determine arc AB, the tooth thickness at the base circle, consider Fig. 7.15, which shows a single tooth and the relevant data.

$$\text{arc AB} = 2 \text{ arc AD}$$
$$= 2 (\text{arc AC} + \text{arc CD})$$

$$\text{arc } \frac{AC}{R_b} = \text{Inv } \psi \text{ radians}$$

$$\therefore \text{ arc AC} = R_b (\tan \psi - \psi)$$

$$\text{arc AC} = \frac{NM}{2} \cos \psi (\tan \psi - \psi) \qquad \ldots (2)$$

$$\theta \text{ radians} = \frac{\text{arc EF}}{R_p} = \frac{\text{arc CD}}{R_b}$$

$$\text{arc EF} = \frac{1}{4} \text{ circular pitch}$$

$$\theta = \frac{\pi M}{4} \times \frac{1}{R_p} = \frac{\pi M}{4} \times \frac{2}{NM}$$

$$\therefore \ \theta = \frac{\pi}{2N} \text{ radians}$$

$$\therefore \text{ arc CD} = R_b \times \theta$$

$$\text{arc CD} = \frac{NM}{2} \cos \psi \times \frac{\pi}{2N} \qquad \ldots (3)$$

But arc AB = 2 (arc AC + arc CD) so substituting from expressions (2) and (3)

$$\text{arc AB} = 2 \left[\frac{NM}{2} \cos \psi (\tan \psi - \psi) + \frac{NM}{2} \cos \psi \frac{(\pi)}{2N} \right]$$

$$= NM \cos \psi \left[\tan \psi - \psi + \frac{\pi}{2N} \right] \qquad \ldots (4)$$

and $W = \text{arc AB} + \text{arc BC}$

Combining (1) and (4)

$$W = NM \cos \psi \left[\tan \psi - \psi + \frac{\pi}{2N} \right] + \left[\frac{2\pi S}{N} \right] \frac{NM}{2} \cos \psi$$

from which $W = NM \cos \psi \left[\tan \psi - \psi + \dfrac{\pi}{2N} + \dfrac{\pi S}{N} \right]$

In this expression, N = number of teeth

M = Module

ψ = pressure angle in radians

S = number of tooth spaces contained in W

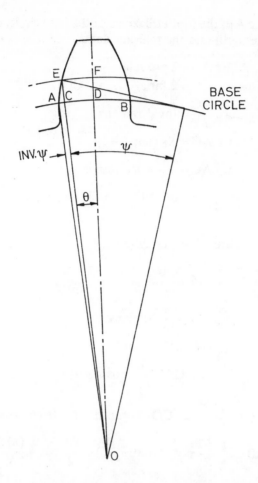

Fig. 7.15. Thickness of a gear
tooth measured at the base circle.

For helical teeth this measurement is made in the normal plane and becomes

$$W_n = NM_n \cos \psi_n \left[\tan \psi_t - \psi_t + \frac{\pi}{2N} + \frac{\pi S}{N} \right]$$

in which $M_n =$ Module in the normal plane

$$= M_t \cos \sigma$$

$\psi_n =$ Pressure angle in the normal plane

Obtained from $\tan \psi_n = \tan \psi_t \cos \sigma$

$\sigma =$ helix angle

Measurements by this method can be improved by the use of either a micrometer with flanged anvils, or the David Brown Base Tangent Comparator shown in Fig. 7.16 which consists essentially of a micrometer with limited movement on either side of a zero setting, the zero setting being made with gauge blocks or distance pieces. Further, the method lends itself to measurement of small tooth gears by projection, either by direct measurement or by traversing the table of a Toolmaker's Microscope.

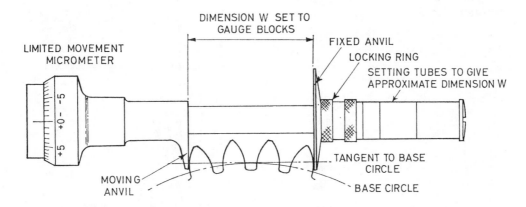

Fig. 7.16.   David Brown Base Tangent Comparator.

Again the measurement W quoted is for uncorrected teeth with no backlash. In this case, if $\delta W_b =$ reduction in W for backlash allowance

$$\delta_b = \text{backlash allowance}$$

then $\delta W_b = - \delta b \cos \psi_n$

and if $K =$ addendum correction

$$\frac{1}{\delta W_c M} = \text{change in } W \text{ due to addendum correction}$$

$$\delta W_c = \pm KM.2 \sin \psi_n$$

Thus, for a corrected gear with backlash,

$$W = NM_n \cos \psi_n \left[\tan \psi_t - \psi_t + \frac{\pi}{2N} + \frac{\pi S}{N} \right] - \delta b \cos \psi_n \pm KM_n 2.\sin \psi n$$

7.11 MEASUREMENT OVER ROLLERS

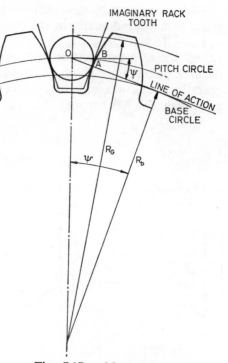

Consider a roller placed in the space between a pair of gear teeth so that its centre lies on the pitch circle as shown in Fig. 7.17.

Working in a similar manner to the constant chord calculation,

$$\cos \psi = \frac{OA}{OB}$$

$$OA = OB \cos \psi$$

But $OA =$ radius of roller r

$$OB = \frac{1}{4}CP = \frac{\pi M}{4}$$

$$\therefore r = \frac{\pi M}{4} \cos \psi$$

and dia. of roller $= \dfrac{\pi M}{2} \cos \psi$

$\therefore$ gauging radius $R_G = R_p + r$

Fig. 7.17. Measurement over a roller whose centre is on the pitch circle.

$$= \frac{NM}{2} + \frac{\pi M}{4} \cos \psi \text{ where } R_p = \text{radius of pitch circle}$$

Thus, the dimension over a pair of rollers in opposite tooth spaces,

$$D_G = NM + \frac{\pi M}{2} \cos \psi = M \left(N + \frac{\pi}{2} \cos \psi\right)$$

For modified teeth the roller radius becomes

$$r \bmod = \frac{\pi M}{4} \cos \psi - 2KM \sin \psi$$

$$\therefore R_G = \frac{NM}{2} + \frac{\pi M}{4} \cos \psi - 2KM \sin \psi$$

$$\text{and } D_G = NM + \frac{\pi M}{2} \cos \psi - 4KM \sin \psi$$

$$= M \left(N + \frac{\pi}{2} \cos \psi - 4K \sin \psi\right)$$

Allowance for backlash is directly added to the roller radius and therefore to R_G.

This case requires a particular sized roller whose radius, $\frac{\pi M}{4}\cos\psi$, is less than the addendum. It will therefore be shrouded by the tips of the teeth and measurement may be difficult. If this is so, readers are directed to the work of Earl Buckingham, who suggests a method using any convenient size of roller.

For gears with an even number of teeth, a direct diameter measurement may be made. If the gear has an odd number of teeth, a radial measurement with the gear between centres can be carried out, using a comparator with the gear, or a measurement over a pair of rollers, provided allowance is made for the angular relationship of the rollers relative to each other.

7.12 GEAR PITCH MEASUREMENT

The measurement of the pitch of gear teeth may be made by (*a*) measuring the distance from a point on one tooth to a suitable point on the next tooth; (*b*) measuring the position of a suitable point on a tooth after the gear has been indexed through a suitable angle.

7.121 Tooth to Tooth Pitch Measurement

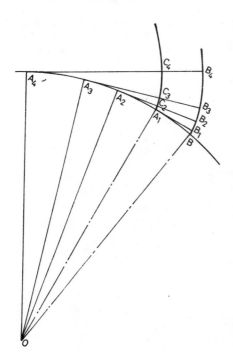

One method of carrying out this measurement is by a portable hand-held instrument which measures the base pitch.

In Fig. 7.18, the base pitch is A_1B between the involutes with origins at A_1 and B.

It can be shown that

$$A_1B = A_1B_1 = C_2B_2 = C_3B_3 = C_4B_4$$

i.e. any measurement along a tangent to the base circle between a pair of adjacent involutes is equal to the base

pitch, and base pitch $P_b = \frac{\pi}{P}\cos\psi$.

The instrument for making this measurement consists of three measuring fingers, as shown in Fig. 7.19.

Fig. 7.18.   The base pitch is equal to the linear distance between a pair of involutes measured along a common generator.

Measuring finger A is located against a tooth so that the mark shown is approximately at the point of tangency.

The instrument is adjusted so that the distance between A and C is approximately equal to the base pitch. The finger B is adjusted to locate the fingers on the same place on each tooth and support the instrument. If the pitch is constant, the reversal reading on each tooth should be the same when the instrument is rocked.

Another possible method is to use two dial gauges on adjacent teeth with the gear mounted in centres, as in Fig. 7.20. The gear is indexed through successive pitches to give a constant reading on dial A. Any changes in the reading on dial B indicate that pitch errors are present. The actual error can be determined by deducting the individual reading on dial B from the mean of the readings.

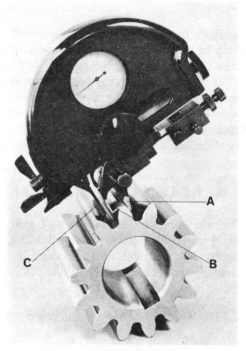

Fig. 7.19.   Maag Pitch Measuring Instrument, Type TMC.
(*Courtesy of the Maag Gear Wheel Co. Ltd., Zürich*)

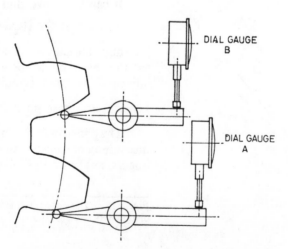

Fig. 7.20.   Use of two dial gauges to determine tooth-to-tooth pitch errors.

150

7.122   Measurement of Cumulative Pitch Error

It would appear that the simplest method of determining pitch errors is to set a dial gauge against a tooth and note the reading. If the gear is now indexed through the angular pitch and the reading differs from the original reading, the difference between these is the cumulative pitch error. The problem is to index through the exact angular pitch, as an error in indexing of $\delta\theta$ radians induces an error of $R_p\delta\theta$ in pitch, where R_p is the pitch circle radius.

However, if the angle through which the gear is indexed is always the same, and not necessarily the angular pitch, the induced error can be corrected. Such an indexing device has been developed at the National Physical Laboratory and is incorporated in an elegant pitch measuring device made by the Sigma Instrument Company.

The indexing device is basically as shown in Fig. 7.21.

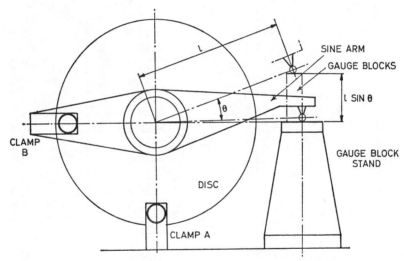

Fig. 7.21.   Indexing device used in determining cumulative pitch errors.

The gear is mounted on a mandrel attached to the disc and rotates with it. Rotating concentrically with the disc is a sine arm. Clamp A locks the disc to the base and clamp B locks the arm to the disc. To index the gear, and starting with clamp A locked, and hence the disc and gear locked together, the procedure is:

(a) With clamp B unlocked, raise the sine arm and place suitable gauge blocks on the stand, such that the arm is rotated through angle θ.

(b) Hold the stylus down on the gauge blocks and lock clamp B.

(c) Release clamp A and remove the gauge blocks.

(d) Swing the sine arm down to the base and lock clamp A.

This process is repeated until readings have been taken on all the teeth and a repeat reading is taken on the first tooth. The difference between the original and final readings is now distributed among the intermediate readings to give the cumulative pitch error on each tooth. An example is shown in the table below, graphs of errors being as in Fig. 7.22.

Tooth Number	Reading (0·01 mm units)	Correction for Induced Error (0·01 mm units)	Cumulative Error (0·01 mm units)
1	0	0	0
2	+ 2·5	− 1·9	+0·6
3	+ 6·0	− 3·8	+2·2
4	+ 8·0	− 5·7	+2·3
5	+11·0	− 7·6	+3·4
6	+13·0	− 9·5	+3·5
7	+14·0	− 11·4	+2·6
8	+14·5	− 13·3	+1·2
9	+15·0	− 15·2	−0·2
10	+16·5	− 17·1	−0·6
11	+17·5	− 19·0	−1·5
12	+18·5	− 20·9	−2·4
13	+21	− 22·8	−1·8
14	+22	− 24·7	−2·7
15	+23	− 26·6	−3·6
16	+26	− 28·5	−2·5
17	+28	− 30·4	−2·4
18	+31	− 32·3	−1·3
19	+36	− 34·2	+1·8
20	+37·5	− 36·1	+1·4
Repeat 1	+38	− 38·0	0

Thus the cumulative pitch error is 7·0 (0·01 mm units) between tooth numbers 5 and 15 over an arc of $10/20 \times 2\pi$ radians.

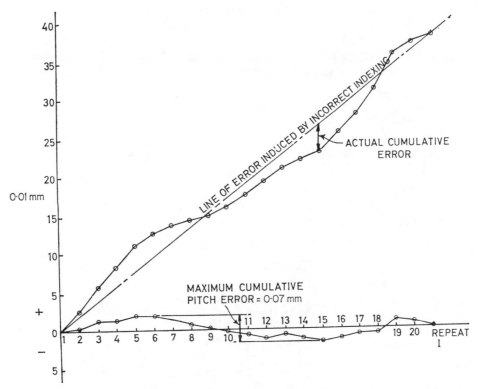

Fig. 7.22.   Graphs of readings and cumulative pitch error. Note that tooth-to-tooth error can be obtained by subtracting adjacent readings.

7.13   TESTING INVOLUTE FORM

If a measuring instrument incorporating a dial indicator gauge is made to trace the path of an involute and its stylus is set against an involute, variations in the involute form would show as variations of the dial gauge reading. The type of instrument outlined in Fig. 7.23 suggests itself as being a possible design. If the straight edge is rolled around the base circle without slipping, the plunger of the dial gauge traverses an involute curve generated from that base circle. If the tooth is of correct involute form, the dial gauge plunger will traverse the tooth being displaced from its original position of zero.

Note that errors are measured normal to the tooth form, i.e. along the line of the straight edge.

This instrument obviously would not be successful, the dial gauge interfering with the next tooth, which is conveniently left off the diagram. However, this is the principle of most involute form testing devices, the gear being mounted on a disc whose diameter is accurately the diameter of the theoretical base circle D_b.

153

$$D_b = D \cos \psi$$
$$= NM \cos \psi$$

The David Brown Involute Form Tester shown in Fig. 7.24 is of this type, the stylus, bearing against the involute form, being accurately located above the

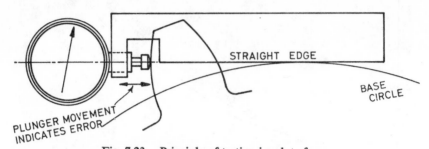

Fig. 7.23. Principle of testing involute form.

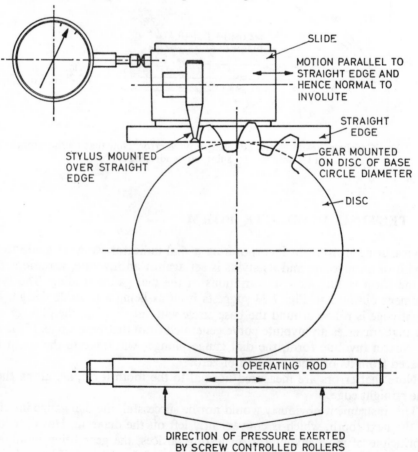

Fig. 7.24. Diagrammatic layout of David Brown Involute Form Tester.

154

straight edge. The stylus is mounted on a small vee/flat ball slide constrained to move parallel to the straight edge. Thus slide movements measured by the dial gauge are in fact errors in involute form measured along the line of the straight edge, i.e. normal to the involute form. Motion without slipping is caused by a lateral movement of the rod at the back of the instrument.

7.14 ALLOWABLE ERRORS IN SPUR GEARS

The allowable errors set out in B.S. 436: 1940 are based on a tolerance factor δ, given by

$$\delta = \left[\frac{(N+60)M}{10} + 1 \right]$$

The only other symbol used which has not been dealt with is concerned with accumulated pitch errors, and is: L_a = length of arc of the pitch circle on which the accumulative error exists.

Referring to section 7.13, it is seen that

$$L_a = \frac{n}{N} \times 2\pi \text{ radians} \times R_p$$

$$= \frac{n}{N} \times 2\pi \times \frac{NM}{2} \quad \text{where } n = \text{number of teeth over which error exists}$$

$$= nM\pi \qquad\qquad\qquad N = \text{number of teeth on gear}$$
$$M = \text{Module}$$

B.S. 436 gives tables of allowable errors in tooth thickness, pitch and profile in terms of δ and L_a. These tables, in association with the calculated values of δ and L_a, enable the actual tolerances to be calculated.

CHAPTER 8

The Measurement of Screw Threads

8.1 INTRODUCTION

EXAMINATION of B.S. 3643, dealing with the tolerances on commercial screw threads, shows that these tolerances are relatively large. Such threads are normally inspected using limit gauges as was described in Chapter 7. However, certain threads must be held to much closer tolerances, and this is particularly true of the limit gauges used for screw thread inspection. These threads must be measured, not gauged, to ensure that they are of a degree of accuracy to separate successfully the good threads from the bad when used as tools of inspection.

Measurement, as distinct from gauging, of a screw thread can be extremely complex. There are a number of elements to be measured and, as will be shown, some are interrelated. A vee-form thread is composed basically of the following elements:

(a) Major or outside diameter.

(b) Minor or root diameter.

(c) Form, particularly flank angles.

(d) Pitch. } Virtual effective diameter.

(e) Simple effective diameter.

These elements are illustrated in Fig. 8.1.

It will be noted that the flank angle, pitch, and simple effective diameter are grouped together under the heading virtual effective diameter. In Chapter 7 it was stated that errors in pitch and/or flank angle caused a change in effective diameter. Thus the virtual effective diameter of a thread is the simple effective diameter modified by corrections due to pitch errors and flank angle errors, and this virtual effective diameter is the most important single dimension of a screw thread gauge.

8.2 MEASUREMENT OF THE MAJOR DIAMETER

The major diameter of a screw thread is defined as the diameter of an imaginary cylinder which contains all points on the crests of the thread.

It is most conveniently measured by means of a bench micrometer (Fig. 8.2).

156

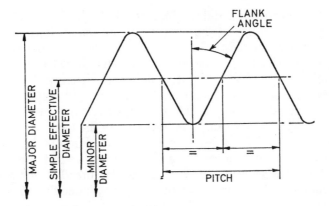

Fig. 8.1. Elements of a vee-form thread.
Note: Metric and Unified threads have flat crests and radiused roots.

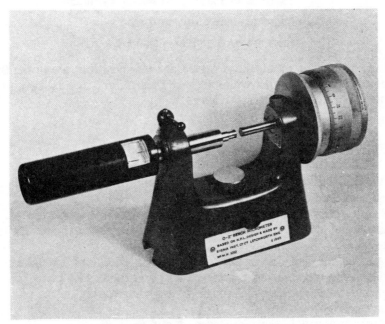

Fig. 8.2. 0–50 mm bench micrometer.
(Courtesy of Herbert Controls and Instruments Ltd.)

This instrument was designed by the National Physical Laboratory to eliminate the deficiencies inherent in the normal hand micrometer. These are

(*a*) Variations in measuring pressure.

(*b*) Pitch errors in the micrometer thread.

The fixed anvil is replaced by a fiducial indicator so that all measurements are made at the same pressure, and this indicator has a positional adjustment which not only increases the range of the instrument but makes it virtually impossible for it to be direct reading. Instead it must be used as a comparator and set to a standard. Thus for a given reading the micrometer thread is used over a short length of travel and any pitch errors it contains are virtually eliminated.

The setting standard may be a gauge block, but for preference a calibrated setting cylinder should be used as this gives greater similarity of contact at the anvils when reading on the setting standard and on the gauge.

The procedure consists simply of noting the reading obtained on the setting cylinder, and that obtained on the thread. The difference in these two readings is then the size difference between the cylinder and the thread at that position. If this difference is added to the diameter of the setting cylinder the result is the major diameter of the thread.

If D_c = calibrated diameter of setting cylinder

R_c = micrometer reading on setting cylinder

R_t = micrometer reading on thread

then major diameter = $D_c + (R_t - R_c)$

This measurement should be repeated at three positions along the thread to determine the amount of taper which may be present, and at least two, preferably three, angular positions to detect ovality.

8.3 MEASUREMENT OF MINOR DIAMETER

The minor diameter may be defined as the diameter of an imaginary cylinder containing all points on the root of the thread.

It is measured by a comparative process similar to that used in section 8.3, but hardened and ground steel prisms are used to probe to the root of the thread as shown in Fig. 8.3.

It can be seen that due to the thread helix a couple of $F \times p/2$ Nm is produced where F is the measuring pressure and p is the thread pitch. This couple would tend to rotate the thread through an angle depending on the pitch, and an erroneous reading would result.

The measurement is therefore carried out on a floating carriage diameter measuring machine in which the thread is mounted between centres and a type of bench micrometer is constrained to move at right angles to the axis of the centres by a vee-ball slide as described in Chapter 3.

The instrument appears as shown in Fig. 8.4.

The readings are taken on the setting cylinder *with the prisms in position*, and on the thread. If the prisms are considered as extensions to the micrometer anvils as in Fig. 8.3 it is seen that their size is unimportant and:

Minor dia. = $D_c + (R_t - R_c)$

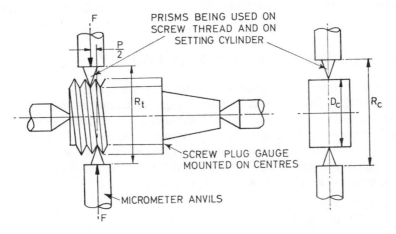

Fig. 8.3.   Use of prisms to measure minor diameter.

Fig. 8.4.   Floating carriage diameter measuring machine.
(*Courtesy of the Coventry Gauge & Tool Co. Ltd.*)

Again readings should be taken at various positions on the thread to determine ovality and taper.

8.4 MEASUREMENT OF THREAD FORM

The most important form measurement to be made on a screw thread is the measurement of its flank angles. The flank angle is defined as the angle made between the straight portion of the thread flank and a line normal to the thread axis.

Flank angles on large threads may be measured by contact methods, but normally the only practical method of measurement is to use optical equipment. This can be by projection and measuring the angle of the flank image on the screen, or by using a microscope with a goniometric head.

8.41 Screw Thread Projection

The simplest and probably the most effective projector for this class of work is known as the N.P.L. projector as it was developed at the National Physical Laboratory.

It consists of a lamp-house whose optical outlet contains condenser lens to give even illumination. The object to be projected is mounted on a stage between the condensers and the projection lens which throws an enlarged image on the screen as in Fig. 8.5.

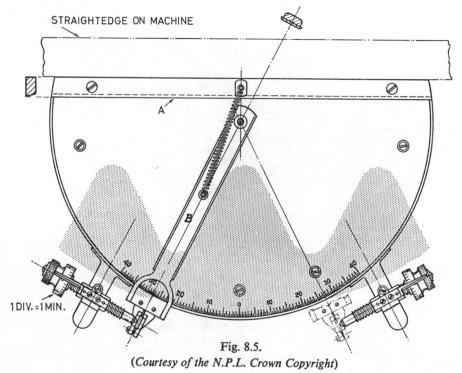

Fig. 8.5.
(Courtesy of the N.P.L. Crown Copyright)

The whole projector may be moved on rails normal to the screen to give the required degree of enlargement, and the work-stage may be moved relative to the projection lens for focusing purposes. This focusing motion can be made from the screen, by an ingenious arrangement of wires and pulleys. Thus an extremely sharp image can be produced at the screen without continual walking from screen to instrument.

An interesting point about the instrument is that to accommodate screw threads the work stage has centres, and can be swung out of normal to the optical axis.

This is to avoid interference due to the helix angle. If the thread is mounted with its axis normal to the optical axis the top of the thread, which is projected, is at an angle to the optical axis due to the thread helix. This is shown exaggerated in Fig. 8.6 (*a*).

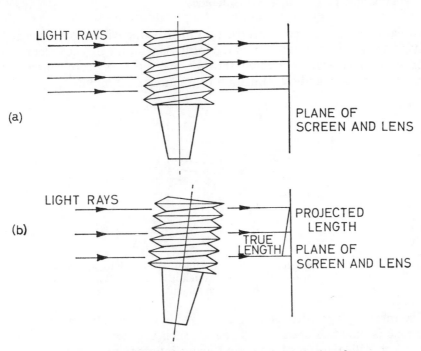

Fig. 8.6(*a*).   Projection normal to thread axis causes interference.
(*b*) Turning thread through helix angle avoids interference but fore-
shortens pitch and distorts profile.

By swinging the thread through its helix angle the thread helix is parallel with the light rays [Fig. 8.6 (*b*)]. However, the flank angle is defined as being measured on a plane section parallel to the thread axis. This being so it can be seen that the set-up in Fig. 8.6 (*b*) has a foreshortening effect and will induce a narrowing of the thread image as projected on the screen.

This problem can be overcome by swinging the lamp-house through the helix angle so that the light rays are not impeded as they pass through the thread to the objective lens.

Note that in this case the lens and screen are parallel to the thread axis and the foreshortening effect, with its consequent distortion of the flank angles, is eliminated.

The actual measurement is carried out using a shadow protractor mounted on a ledge on the screen. The angle of the ledge can be adjusted until it is parallel with the image of the crests or roots of the thread. It is then assumed to be parallel to the thread axis and is used as a datum or base for the measurement.

The shadow protractor and image set up for measurement are shown in Fig. 8.5.

8.42 Microscopic Flank Angle Measurement

Thread flank angles may be measured by a microscope with a goniometric head. This consists of a clear glass screen in the focal plane of the objective lens carrying datum lines which can be rotated through 360°, the angle of rotation being measured direct to 1′ and by estimation to fractions of a minute.

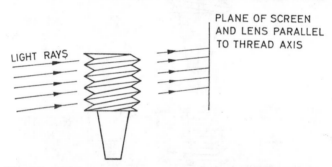

Fig. 8.7. With thread axis and screen parallel, interference is avoided by turning light rays through helix angle. Distortion is kept to a minimum.

The thread gauge is mounted on centres and illuminated from below. The microscope is mounted above the thread in such a way that it can be swivelled to be in line with the thread helix and avoid interference of the image. This is shown in diagram (Fig. 8.8).

The centres are mounted on slideways which enable them to be moved through co-ordinate dimensions by micrometers reading to 0·002 mm and on a rotary table.

In operation the microscope is focused with its axis vertical, on a focusing bar set so that it is focused on a plane through the line of the centres, i.e. the axis of the thread to be measured. The thread is then set up in place of the focusing bar and the microscope swung through the helix angle of the thread to avoid interference.

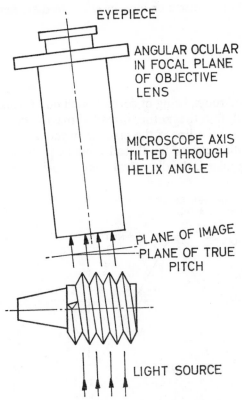

EYEPIECE

ANGULAR OCULAR
IN FOCAL PLANE
OF OBJECTIVE
LENS

MICROSCOPE AXIS
TILTED THROUGH
HELIX ANGLE

PLANE OF IMAGE
PLANE OF TRUE
PITCH

LIGHT SOURCE

Fig. 8.8.   Microscopic measurement of flank
angles.

The datum lines in the microscope head are set to zero and the table rotated until the crests of the thread image coincide with the horizontal datum. The table is then locked and the datum lines in the microscope eyepiece rotated until they coincide with the thread flanks. The flank angles are then read off the eyepiece scale.

This equipment is normally among the attachments which may be set up on a toolmaker's microscope. The readings and settings can either be made through the microscope eyepiece or viewed on a screen as shown in Fig. 8.9.

It should be noted that the optical axis is not normal to the thread axis and some distortion of the image still occurs. However, this technique is quite easy to set up and produces results of a reasonably high order of accuracy.

Whichever method is used, errors on both right- and left-hand flanks should be determined as each can cause interference with the mating

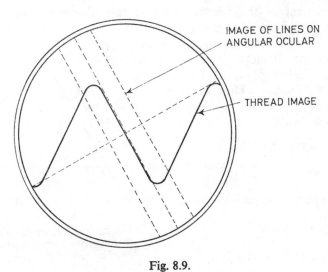

IMAGE OF LINES ON
ANGULAR OCULAR

THREAD IMAGE

Fig. 8.9.

163

thread independently of the other, and both cause a change in effective diameter of the thread under test.

8.43   Effect of Flank Angle Errors

Consider a vee thread having flat crests and roots, being of correct pitch but having an error on one flank only. The effect is to foul the mating thread above the pitch line if the error is positive and below the pitch line if the error is negative as in Fig. 8.10 (*a*). In either case the effective diameter of the thread must be increased for an external thread if fouling is to be avoided, the diametral increase being δE_d.

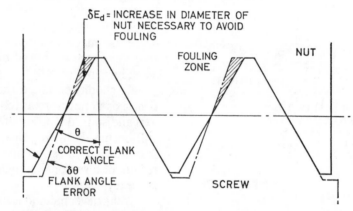

Fig. 8.10(*a*).   Nut of perfect form mating with a screw having a flank angle error $\delta\theta$ on one flank only.

An enlargement of the relevant part of this diagram enables δE_d to be determined. It should be noted that apparently the diametral increase is $2\delta E_d$, but this is not so, because as the diameter is increased an axial movement takes place if the opposite flanks remain in contact. Thus the apparent increase is halved and the net increase in effective diameter is δE_d.

Referring to Fig. 8.10 (*b*):

$$\frac{GD}{ED}=\sin\theta \ \therefore\ ED=\frac{GD}{\sin\theta}$$

If $\delta\theta$ is small and in radians then

$$\frac{GD}{AD}=\frac{GD}{AC}=\delta\theta$$

$$\therefore\ GD=AC\delta\theta \ \text{and} \ ED=\frac{\delta\theta}{\sin\theta}$$

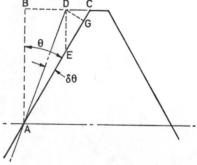

Fig. 8.10(*b*).   Enlarged view of fouling with a screw having a flank angle error $\delta\theta$. The virtual increase in effective dia. $\delta E_d = DE$.

But
$$\frac{AB}{AC} = \cos\theta \therefore AC = \frac{AB}{\cos\theta}$$

But AB is half the depth of thread h

$$\therefore ED = \frac{h}{2}\frac{\delta\theta}{\cos\theta\sin\theta} = \frac{h\delta\theta}{\sin 2\theta}$$

in which θ = nominal flank angle

$\delta\theta$ = flank angle error in radians

h = depth of thread considering the straight portion of the flank only.

Making a similar correction for the opposite flank the expression becomes

$$\delta E_d = \frac{h}{\sin 2\theta}(\delta\theta_1 + \delta\theta_2)$$

where $\delta\theta_1$ and $\delta\theta_2$ are the worst flank angle errors on the r.h. and l.h. flanks in radians.

Applying this expression to a Metric thread form, for which the value for h is given in terms of pitch, from Fig. 8.11; θ is known, and $\delta\theta_1$ and $\delta\theta_2$ may be converted from radians to degrees.

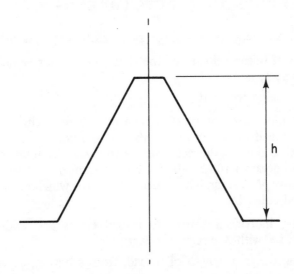

Fig. 8.11. Metric thread form in which h = 0·54127p.

From Fig. 8.11:

$$h = 0·54127p$$

If this value for h is substituted in the expression for change in effective diameter it becomes

$$\delta E_d = \frac{0·5413p}{\sin 2\theta}(\delta\theta_1 + \delta\theta_2)$$

165

in which θ is $30°$, the flank angle, and $\delta\theta_1$ and $\delta\theta_2$ are in radians and must be changed to degrees.

$$\therefore \; \delta E_d = \frac{0 \cdot 5413p}{0 \cdot 8667}(\delta\theta_1 + \delta\theta_2) \times \frac{2\pi}{360}$$

$$\delta E_d = 0 \cdot 0109p \; (\delta\theta_1 + \delta\theta_2)$$

In practice the form of a metric thread may be modified by having root radius giving a greater depth of straight flank than $0 \cdot 5413p$ and to allow for this a value of $0 \cdot 0115p(\delta\theta_1 + \delta\theta_2)$ is generally used.

Similar expressions can be derived for other thread forms as follows:

B.A. threads $(47\frac{1}{2}°)$ $\delta E_d = 0 \cdot 0091p \; (\delta\theta_1 + \delta\theta_2)$

Unified threads $(60°)$ $\delta E_d = 0 \cdot 0098p \; (\delta\theta_1 + \delta\theta_2)$

Whitworth threads $(55°)$ $\delta E_d = 0 \cdot 0105p \; (\delta\theta_1 + \delta\theta_2)$

It must be recalled that the values of $\delta\theta_1$ and $\delta\theta_2$ are the flank angle errors in *degrees*, and p is the nominal pitch of the thread being measured.

8.5 PITCH ERRORS IN SCREW THREADS

If a screw thread is generated by a single point cutting tool its pitch depends on:

(*a*) the ratio of linear velocity of the tool and angular velocity of the work being correct;

(*b*) this ratio being constant.

If these conditions are not satisfied then pitch errors will occur, the type of error being determined by which of the above conditions is not satisfied. Whatever type of error is present the net result is to cause the total length of thread engaged to be too great or too small and this error in overall length of thread is called the *cumulative pitch error*. This, then, is the error which must be determined. It can be obtained either by:

(*a*) measuring individual thread to thread errors and adding them algebraically, i.e. with due regard to sign;

(*b*) measuring the total length of thread, from a datum, at each thread and subtracting from the nominal value.

8.51 Types of Pitch Error

8.511 *Progressive Pitch Error*

This error occurs when the tool–work velocity ratio is constant but incorrect. It may be caused through using an incorrect gear train, or an approximate gear train between work and tool lead screw as when producing a metric thread with an inch pitch lead screw when no translatory gear is available. More commonly, it is caused by pitch errors in the lead screw of the lathe or other generating machine.

If the pitch error per thread is δp then at any position along the thread the cumulative pitch error is $n\delta p$ where n is the number of threads considered. A graph of cumulative pitch error against length of thread is therefore a straight line [Fig. 8.12 (*a*)].

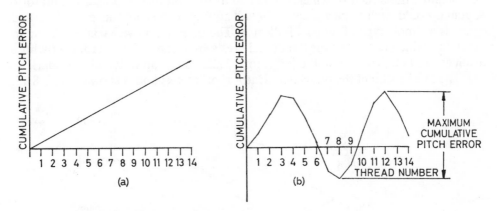

Fig. 8.12(*a*).   Progressive pitch error. (*b*) Periodic pitch error.

8.512   *Periodic Pitch Error*

This type of error occurs when the tool–work velocity ratio is not constant. It may be caused by pitch errors in the gears connecting the work and lead screw or by an axial movement of the lead screw due to worn thrust faces. Such a movement would be superimposed on the normal tool motion to be reproduced on the work. It will be appreciated that errors due to these causes will be cyclic, i.e. the pitch will increase to a maximum, reduce through normal to a minimum and so on.

A graph of cumulative pitch error will thus be of approximately sinusoidal form as in Fig. 8.12 (*b*), and the maximum cumulative pitch error will be the total error between the greatest positive and negative peaks within the length of thread engaged.

8.513   *Thread Drunkenness*

A drunken thread is a particular case of a periodic pitch error recurring at intervals of one pitch. This means that the pitch measured parallel to the thread axis will always be correct, and all that is in fact happening is that the thread is

167

not cut to a true helix. A development of the thread helix will be a curve and not a straight line. Such errors are extremely difficult to determine and except on large threads will not have any great effect.

8.52 Measurement of Pitch Error

Apart from drunken threads, pitch errors may be determined using a pitch measuring machine, the design of which originated at the National Physical Laboratory. A round-nosed stylus engages the thread approximately at the pitch line and operates a simple type of fiducial indicator. The thread is moved axially relative to the stylus, which can ride over the thread crests by means of a micrometer whose readings are noted each time the indicator needle comes up to its fiducial mark.

The mechanism of the fiducial indicator is of interest and is shown in Fig. 8.13.

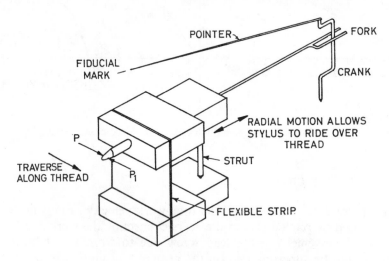

Fig. 8.13.   Fiducial indicator used on pitch-measuring machine.

The stylus is mounted in a block supported by a thin metal strip and a strut. It may thus move back and forth over the threads, the strut and strip giving a parallel-type motion. If, however, the side pressures on the stylus, P and P_1, are unequal the strip twists and the block pivots about the strut. The forked arm causes the crank to rotate and with it the pointer. Thus the pointer will only be against the fiducial mark when the pressures P and P_1 caused by the stylus bearing on the thread flanks are the same in each thread.

Errors in the micrometer are reduced by a cam-type correction bar and, with care, accuracies of greater than 0·002 mm may be consistently achieved.

If thread to thread pitches are required then each micrometer reading is subtracted from the next. More usually cumulative pitch errors are required and can be obtained by simply noting the micrometer readings and subtracting them from the expected reading. This should normally be repeated with the thread turned through 180° in case the thread axis does not coincide with the axis of the centres on which it is mounted. The mean of the two readings, usually determined graphically, is then used as the pitch error.

8.53 Effects of Pitch Errors

If a thread has a pitch error it will only enter a nut of perfect form and pitch if the nut is made oversize. This is true whether the pitch error is positive or negative, and thus, whatever pitch error is present in a screw plug gauge, it will reject work which is near the low limit of size.

Consider a thread having a cumulative pitch error of δp over a number of threads, i.e. its length is $np + \delta p$. If such a screw is engaged with a nut of perfect form and pitch they will mate as shown in Fig. 8.14 (*a*).

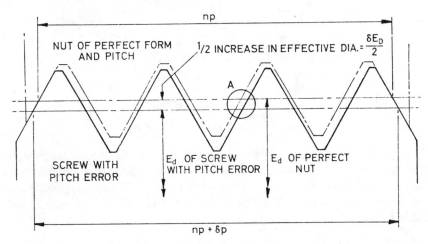

Fig. 8.14(*a*). Screw having cumulative pitch error δp in mesh with a nut of perfect form and pitch.

Consider an enlarged view of the thread flanks at A as in Fig. 8.14 (*b*). It is seen that

$$\tan \theta = \frac{\dfrac{\delta p}{2}}{\dfrac{\delta E_d}{2}}$$

$$= \frac{\delta p}{\delta E_d}$$

$$\therefore \quad \delta E_d = \delta p \cotan \theta$$

where δp is the cumulative pitch error over the length of engagement and δE_d is the equivalent increase in effective diameter

169

The importance of this is emphasized when a Whitworth thread is considered in which the flank angle θ is $27\frac{1}{2}°$ and cotangent $27\frac{1}{2}° = 1 \cdot 920$.

$$\text{For Whitworth threads } \delta E_d = 1 \cdot 920 \; \delta p$$
$$\text{For Metric threads } \delta E_d = 1 \cdot 732 \; \delta p$$

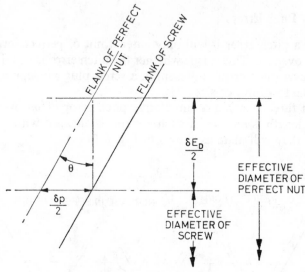

Fig. 8.14(*b*).   Enlarged view at *A*.

The pitch error is therefore almost doubled when the equivalent increase in effective diameter is calculated. A screw plug gauge having a cumulative pitch error of 0·006 mm will thus reject all work within 0·012 mm (approximately) of the low limit in the case of Whitworth threads, and within 0·01 mm in the case of Metric threads.

8.6 MEASUREMENT OF SIMPLE EFFECTIVE DIAMETER

The simple effective diameter of a screw thread may be defined as the diameter of an imaginary cylinder co-axial with the thread axis which cuts the thread so that the distance between any pair of intercepts and adjacent flanks is half the pitch. This is explained more simply diagrammatically in Fig. 8.15.

Measurement of the simple effective diameter is carried out on the machine described in section 8.2, the prisms used for measuring minor diameter being replaced by steel wires, or cylinders, whose size is chosen so that they 'pitch' approximately at the effective diameter. Such cylinders are known as 'Best Size' wires and enable nominal values of pitch and flank angle to be used in subsequent calculations. Such cylinders may be purchased from most manufacturers of gauging equipment and their size is calibrated.

As for the minor diameter, a reading is taken *with the cylinders*, over a setting

170

cylinder. A reading is then taken with the cylinders engaged in the thread. The difference in the readings is the difference between the diameter of the setting cylinder and the dimension *T under* the wires when engaged with the thread.

The geometry of this arrangement is shown in Fig. 8.15, and is used in deriving the expression required to calculate the simple effective diameter.

From the diagram (Fig. 8.15):

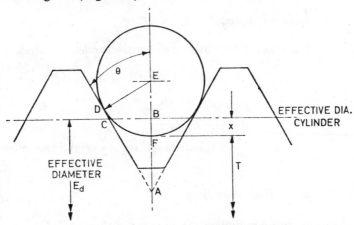

Fig. 8.15. Calculation of simple effective diameter.

$$E_d = T + 2x$$

In $\triangle$ ABC $\tan \theta = \dfrac{BC}{AB}$

$$\therefore \ AB = BC \cotan \theta$$

But by definition of the simple effective diameter, $BC = \frac{1}{4}$ pitch

$$\therefore \ AB = \frac{p}{4} \cotan \theta \qquad \qquad \dots (1)$$

In $\triangle$ ADE $\sin \theta = \dfrac{DE}{AE}$

$$\therefore \ AE = DE \cosec \theta$$

But $\qquad DE = \frac{1}{2}$ diameter of the measuring wires

$$\therefore \ AE = \frac{d}{2} \cosec \theta$$

Now $\qquad x = AB - AF$

and $AF = AE - EF$

but EF also $= \dfrac{d}{2}$

$$\therefore \ AF = \frac{d}{2} \cosec \theta - \frac{d}{2}$$

$$AF = \frac{d}{2} (\cosec \theta - 1) \qquad \qquad \dots (2)$$

171

Subtracting (2) from (1) to obtain x we get

$$x = \frac{p}{4} \cotan \theta - \frac{d}{2} (\cosec \theta - 1)$$

But $E_d = T + 2x$ and the term $2x$ is constant for any given thread if the nominal values of pitch and flank angle are used, as they may be if the best wire size is used.

$$\therefore \quad P = 2x = \frac{p}{2} \cot \theta - (\cosec \theta - 1)d$$

$$\therefore \quad E_d = T + P \text{ in which } T = \text{measured dimension } under \text{ the wires}$$

$$P = \frac{p}{2} \cot \theta - (\cosec \theta - 1)d$$

$$p = \text{nominal pitch}$$

$$d = \text{wire diameter}$$

$$\theta = \text{nominal flank angle} = \text{semi-angle of thread}$$

8.61 Calculation of the Best Wire Size

Normally when measuring cylinders are purchased they are of the best wire size for a given thread. However if they are not available it is best to make them rather than use incorrect wire sizes as this may incur large diametral errors due to flank angle errors.

In Fig 8.16 AB is $\frac{1}{2}$ wire diameter and BC is $\frac{1}{4}$ pitch.

$$\cos \theta = \frac{BC}{AB}$$

$$AB = \frac{BC}{\cos \theta}$$

$$\frac{d}{2} = \frac{p}{4 \cos \theta}$$

$$d = \frac{p}{2 \cos \theta}$$

Note. Study of this simple diagram may suggest a simple method of determining the simple effective diameter. This must *not* be used as the wire will rarely pitch on the effective diameter. The discrepancy will not affect the previous method sensibly if flank angle errors are present, but will have considerable effect if the incorrect simplified method is used.

8.62 Measurement over the Wires

If a measuring machine is not available the simple effective diameter may be

172

determined by measuring *over* three wires, or a better method by the use of 'Ovee' gauges, as shown in Fig. 8.17 which may be sprung over the thread.

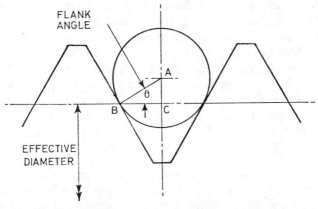

Fig. 8.16. 'Best size' cylinder contacting thread at effective diameter.

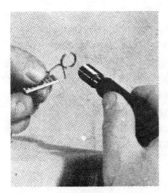 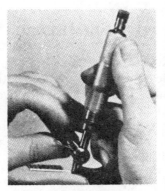 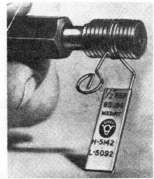

INSERT	MEASURE	COMPARE
Keep coil expanded by depressing lever and screw in component. Avoid friction.	Take reading over coils.	Check reading with limits on data plate.

Fig. 8.17. Use of Ovee spring gauges.
(*Courtesy of the O-Vee Spring Gauges Ltd.*)

If the dimension T_1 *over* the wires is determined then:

$$E_d = T_1 + P_1 = T + P$$
$$\text{and } T_1 = T + 2d$$

$$\therefore P_1 = P - 2d = \frac{p}{2} \cot \theta - d (\operatorname{cosec} \theta - 1) - 2d$$

$$\text{or } P_1 = \frac{p}{2} \cot \theta - d (\operatorname{cosec} \theta + 1)$$

173

8.63   Corrections for Raking and Elastic Compression

The expression for simple effective diameter derived in section 8.6 is based on the assumption that the measurement is made over a plane section of the thread parallel to the thread axis, i.e. measurement is over a series of annular grooves of thread form. This of course is not so and the effect of measuring over a helix is to throw the measuring cylinders out slightly and increase the measured value by an amount C. At the same time measuring pressure applied to point contacts causes elastic deformation of the wires and the thread flanks and reduces the measured value by an amount e. The correct value is obtained by applying corrections of $-C$ and $+e$.

$$\therefore E_d = T + P - C + e$$

The magnitudes of C and e are both normally very small, and as they are of opposite sign their total effect is still less and they may be disregarded. If it is required to determine their values for the measurement of reference gauges, readers are referred to the National Physical Laboratory Notes on Applied Science II, 'Measurement of Screw Threads', obtainable from H.M. Stationery Office.

8.7   VIRTUAL EFFECTIVE DIAMETER

From sections 8.4 and 8.5 it is seen that the effect of pitch and flank angle errors on a screw plug gauge is to cause a virtual increase in the effective diameter in the case of external screw threads. Thus the virtual effective diameter is defined as the effective diameter of the smallest nut of perfect form and pitch which the screw will enter. This is of course the vital dimension of a screw plug gauge and it is obtained by adding to the simple effective diameter amounts depending on the magnitude of the pitch and flank angle errors.

$\therefore$ Virtual effective diameter $=$ Simple effective diameter $+ K_1 p\,(\delta\theta_1 + \delta\theta_2) + \delta p \cot \theta$

in which $K_1 =$ a constant depending on the thread form

$p =$ thread pitch

$\theta =$ thread flank angle

$\delta p =$ pitch error over length of engagement

For a Whitworth thread
virtual effective diameter $=$ simple effective diameter $+ 0 \cdot 0105\, p(\delta\theta_1 + \delta\theta_2) + 1 \cdot 920 \delta p$
and *for a Unified thread*
virtual effective diameter $=$ simple effective diameter $+ 0 \cdot 0098 p(\delta\theta_1 + \delta\theta_2) + 1 \cdot 73\ \delta p$
and *for a Metric thread*
virtual effective diameter $=$ simple effective diameter $+ 0 \cdot 0115 p(\delta\theta_1 + \delta\theta_2) + 1 \cdot 732\ \delta p$

174

8.8 MEASUREMENT OF SCREW RING GAUGES

The problems of measuring screw ring gauges are essentially the same as for plug gauges but the difficulty is increased by the inaccessibility of the thread form and dimensions.

The pitch may be measured on a normal screw pitch measuring machine using a special attachment to allow entry of the stylus, rather like a boring bar.

The flank angle may only be measured by making a plaster cast of the thread, which must be of less than half the diameter of the thread and *lifted out*—not screwed out or the cast may be distorted. This may then be projected in the normal way.

The simple effective diameter is measured using ball-ended stylii, of best wire size, in a measuring machine, the machine being set to a master consisting of a pair of vee notched arms analogizing the thread (i.e. the master is built up so that the effective diameter of the vees is the nominal value for the thread to be measured).

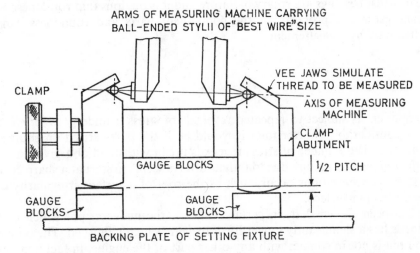

Fig. 8.18. Setting a measuring machine for determining simple effective diameter of an internal thread. Note the vee jaws offset by half pitch and the floating work stage allowing the fixture to swing into the attitude taken up by the thread.

These methods may only be used for fairly large threads. The only way with very small screw ring gauges is to procure very accurate master screw plug gauges whose sizes differ by the gauge tolerance, and use them to test the gauge on a 'GO' or 'NOT GO' basis.

It should be noted that the effect of pitch and flank angle errors is opposite on a screw ring gauge (i.e. these errors tend to *reduce* the simple effective diameter and the corrections for these errors must be subtracted and not added).

CHAPTER 9

Measurement of Surface Texture and Roundness

9.1 INTRODUCTION

THE development of modern technologies has called not only for improved control of dimensional accuracy but also of the texture and geometric form of both working and non-working surfaces of components. There are three main factors which have made the control of surface texture important: fatigue life, bearing properties and wear. Most bearings are rotary in nature and it is obvious that roundness is an important geometric property of the component parts and roundness can be controlled only by measurement.

9.11 Fatigue Life

If a component is subject to repeated reversals of stress it undergoes *fatigue* and its life is considerably shorter than it would be if the part carried an equivalent constant load. The number of stress reversals it can withstand at a given stress is called the *fatigue life*. Failure due to fatigue always seems to start at a sharp corner, where stress concentrations occur, such as the root of a surface irregularity even on a non-working surface.

It has been shown that, in certain engines, the gudgeon pin which carries cyclic loads has a significantly longer life if its *bore* is highly finished. The bore of a gudgeon pin is not in contact with any other part of the engine, in fact a gudgeon pin is usually hollow simply to reduce the reciprocating weight.

9.12 Bearing Properties

A 'perfect' surface, i.e. one with no irregularities whatsoever and therefore 'perfectly' smooth is not a good bearing. In fact seizure would probably occur due to difficulty of maintaining a lubricating film and presenting metal to metal contact. Probably the best form of surface for a bearing is one similar to Fig. 9.1 (*b*) on page 177, in which the large contact areas reduce friction, and the valleys help to retain a film of lubricant.

176

9.13 Wear

It is a well-known law of physics that friction does not depend on contact area. However, the rate of wear is dependent on the areas in contact, the larger the area the lower the load per unit area and hence the lower the rate of wear.

9.2 THE MEANING OF SURFACE TEXTURE

Before anything can be measured it is necessary to define what is to be measured. In most cases of measurement this is not too difficult but in the case of surface texture the definition is not so easy.

Consider a cricket field. The requirements of the surface of the table on which the wickets are pitched are that it shall be flat and smooth. Now a field of pasture

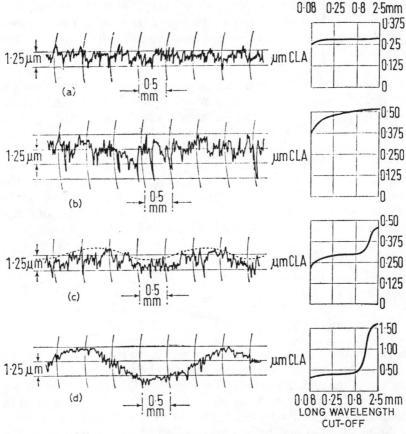

Fig. 9.1. Effect of wavelength cut-off on numerical assessment of surface finish.

(*Courtesy of the Rank Taylor–Hobson Division and R.E. Reason*)

may be undulating but quite smooth, i.e. it may have a series of waves across it say 20 cm deep and 3 m from crest to crest. Although smooth it is not flat. It may be decided that if the field was ploughed the resultant earth movements would remove the undulation and so it is ploughed with furrows 20 cm deep but only 30 cm from crest to crest. If the work is successful the field will now be *flat* but still unsuitable for cricket because it is *rough*, and yet the height of the irregularities is the same—all that has changed is the peak to peak distance or the wavelength.

If such an experiment was carried out it would probably make matters worse as the field would finally be both undulating and rough, i.e. the roughness would be superimposed on the waviness, and the effort expended would have been better spent on either smoothing a rough field or flattening a wavy one. However, the point is that the difference between roughness and waviness is one of wavelength rather than depth of irregularities and at some point it must be decided which changes in elevation constitute roughness and which constitute waviness.

Usually waviness is considered an error of form due to incorrect geometry of the process producing the surface and as such is outside the field of surface texture. Roughness may be defined as the irregularities which are an inevitable consequence of the process if carried out on a geometrically perfect machine and it is the isolation and measurement of these irregularities with which this work is concerned.

The manner in which the two may be superimposed on a machined part can be illustrated by considering the requirements of a lathe to produce a truly cylindrical component. It is simply necessary that the tool shall move parallel to the axis of rotation. Consider a lathe in which one centre is higher than the other. This would cause a bobbin shaped surface (hyperboloid of revolution) to be generated. Further, if the tool motion is not controlled in a straight line a series of shorter waves may be superimposed on the long wave curve, and over all of this will be superimposed the feed marks of the tool. A still shorter wavelength may be due to the tool chattering and the surface thus contains four types of irregularity (regardless of any non-roundness) causing it not to be a true cylinder. These are:

(*a*) Due to misalignment of centres. ⎱
(*b*) Due to non-linear feed motion. ⎰ Waviness.

(*c*) Due to tool feed rate. ⎱
(*d*) Due to tool chatter. ⎰ Roughness.

One of the problems in measuring surface finish is to separate the waviness from the roughness and the way this is done may be understood by reverting to a consideration of the undulating field. If we view the field as a whole the waviness is predominant. If, however, we limit our examination to a 30 cm length of the field where undulations have a wavelength of 3 m we should include in our sample little of the undulations and be able to concentrate on the roughness. Similarly if we limit our examination of a machined surface to a length which excludes long wavelengths then only the roughness will be included. The maximum wavelength

Note: Prefix K indicates that a wavelength cut-off of 0·8 mm has been used.

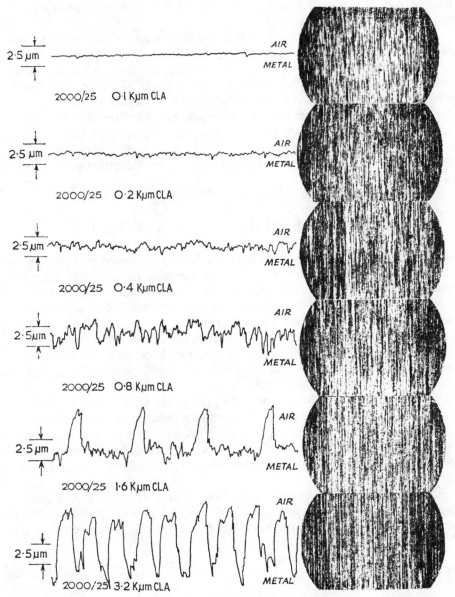

Fig. 9.2.   Surfaces of similar appearance but having vastly different characteristics.
(*Courtesy of the Rank Taylor–Hobson Division and R.E. Reason*)

considered is known as the *cut-off wavelength* which is standardized in Great Britain by B.S. 1134 at 0·25 mm, 0·8 mm, and 2·5 mm; that selected being dependent on the surface being checked.

If a surface contains no waviness then the same value of surface finish will be obtained whatever length is sampled, as shown in Fig. 9.1 (*a*). As the waviness increases the numerical value of the surface texture increases as the cut-off wavelength is increased, as shown in Figs. 9.1 (*b*), (*c*), and (*d*) on p. 177. These all represent ground specimens with varying degrees of chatter marks.

For ground surfaces a cut-off, or sampling length, of 0·8 mm is usual. For short parts such as piston rings 0·25 mm is more suitable as on such short parts there cannot be a succession of events lying far apart. A general rule for a surface produced by a single point tool is that the cut-off wavelength should exceed the feed spacing.

9.3 METHODS OF MEASURING SURFACE FINISH

The methods available for measuring the finish of a machined part may be comparative or by direct measurement. The comparative methods are attempts to assess the surface texture by observation or feel of the surface. They are necessarily comparative in that they can be misleading if comparison is not made with surfaces produced by similar techniques. The appearance of a surface depends largely on the scratch pattern, in the direction of the scratches, and is less influenced by their depths. This is equally true of microscopic examination as is shown in Fig. 9.2 which represents microscopic enlargements of six different surfaces, and their surface profile as shown by the traces.

Touch is probably a better method of assessing surface texture than visual observation, but again it can be misleading and comparisons should only be made of surfaces produced by similar processes. Comparison standards are available, those made by Messrs. Rubert & Co. of Manchester being for different machining processes, while the Norton Grinding Wheel Co. produce cylindrical ground standards.

Direct measurement methods have been developed to enable a numerical value to be placed on the surface finish and these are almost all stylus probe type instruments although interferometric methods are suitable for reflective surfaces.

9.31 Stylus Probe Instruments

In all cases instruments of this type can be broken down into the following units:

(*a*) A skid or shoe drawn slowly over the surface and following its general contours, thus providing a datum.

(*b*) A stylus or probe which moves over the surface with the skid, and vertically relative to the skid due to the roughness of the surface.

(c) An amplifying device for magnifying the stylus movements.

(d) A recording device to produce a trace or record of the surface profile. It should be noted that all such traces are distorted, i.e. the vertical and horizontal magnifications differ, to enable significant vertical features to be observed on a trace of reasonable length. If a sample length of 0·8 mm was magnified 5000 times the resulting trace would need to be 4 m long!

(e) A means of analysing the profile thus obtained. This may be incorporated in the instrument or done separately. Instruments satisfying these characteristics can be produced with mechanical or electronic systems of magnification.

9.32   The Tomlinson Surface Meter

This instrument uses mechano-optical magnification methods and was designed by Dr. Tomlinson of the National Physical Laboratory, its essentials being shown in Fig. 9.3.

The shoe is attached to the body of the instrument, its heights being adjustable to enable the diamond shown to be positioned conveniently. The stylus is restrained from all motion except a vertical one by a leaf spring and a coil spring, the tension in the coil spring P causes a similar tension in the leaf spring. These forces hold a cross roller in position between the stylus and a pair of parallel fixed rollers as shown in the plan view. Attached to the cross roller is a light spring steel arm carrying at its tip a diamond which bears against a smoked glass screen.

In operation the body of the instrument is drawn slowly across the surface by a screw turned at 1 rev/min by a synchronous motor, the glass remaining stationary. Irregularities in the surface cause a vertical movement of the stylus which causes the cross roller to pivot about point A and thus produce a magnified motion of the diamond on the arm. This motion, coupled with the horizontal movement, produces a trace on the glass magnified in the vertical direction at × 100, there being no horizontal magnification.

The smoked glass is transferred to an optical projector and magnified further at × 50 giving an overall vertical magnification of × 5000 and a horizontal magnification of × 50. The trace may be taken off by hand or by photographic methods and analysed.

9.33   The Taylor–Hobson 'Talysurf'

The Talysurf is an electronic instrument whose function can be broken down in a similar manner to the Tomlinson instrument and differing mainly in the method of magnification. It gives the same information much more rapidly and probably more accurately. It can be, and is, used on the factory floor or in the laboratory. The Tomlinson instrument is essentially for use in the laboratory.

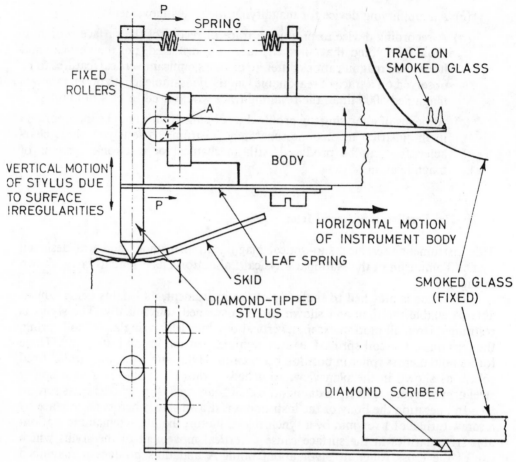

Fig. 9.3. Tomlinson Surface Meter.

The measuring head, as on the Tomlinson, consists of a stylus and shoe (T) which are drawn across the surface under test by an electric motor and gearbox. In this case the arm carrying the stylus forms an armature which pivots about the centre piece of a stack of E shaped stampings around each of the outer pole pieces of which is a coil carrying an a.c. current as in Fig. 9.4.

As the armature pivots about point M it causes the air gaps to vary and thus modulates the amplitude of the original a.c. current flowing in the coils. As these form part of a bridge circuit the output consists of the modulation only. This is fed to an amplifier and caused to operate a pen recorder to produce a permanent record, and to a meter to give a numerical assessment direct.

The pen recorder is of interest in that the trace is produced on carbon-backed paper by a pointer whose tip arcs electrically across through the paper. This produces a much finer line than ink type pen recorders, with no distortion due to drag.

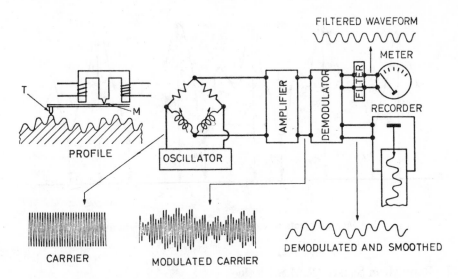

Fig. 9.4. Schematic layout of the Taylor–Hobson 'Talysurf'.
(*Courtesy of the Rank Taylor–Hobson Division and R.E. Reason*)

9.4 ANALYSIS OF SURFACE TRACES

A numerical assessment may be assigned to a surface to indicate its degree of smoothness (or roughness), in a number of ways. Different countries use different techniques, for instance Great Britain, many Commonwealth countries, and the U.S.A. (since 1955), use an average height; Sweden and many continental European countries use the peak to valley height value, while other countries use a root mean square value.

9.41 Peak to Valley Height

The name peak to valley height would indicate a relatively simple method of analysis, but if interpreted in its wider sense it means that the peak and valley used would almost certainly be exceptional and the value obtained would not give a representative assessment of the surface.

To overcome this lack of representation the *ten-point height* average (R_z) is used. This is determined by drawing a line AA parallel to the general lay of the trace, as shown in Fig. 9.5. The heights from AA to the five highest peaks and the five lowest valleys in the trace are determined. The average peak to valley height, R_z, is then given by:

$$R_z = \frac{(h_1 + h_3 + h_5 + h_7 + h_9) - (h_2 + h_4 + h_6 + h_8 + h_{10})}{5} \times \frac{10^3}{\text{vertical magnification}} \ \mu\text{m}.$$

183

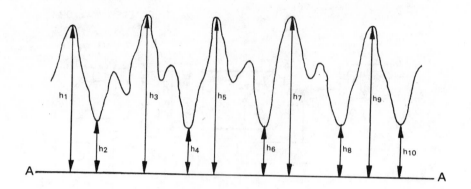

Fig. 9.5. Measurements to calculate the ten-point height average (R_z).

9.42 Root Mean Square (R.M.S.) Value

This measure was standard until 1955 in the U.S.A. after which the standard was changed to an average height (see section 9.43). It is defined as the square root of the mean of the squares of the ordinates of the surface measured from a mean line.

Consider Fig. 9.6.

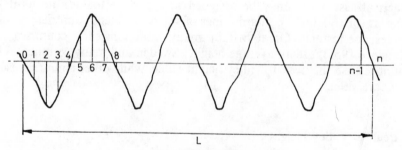

Fig. 9.6. Graphical representation of $h_{r.m.s.}$

If equally spaced ordinates are erected at 1, 2, 3, 4, ... n, whose heights are h_1, h_2, h_3, ... h_n, then

$$h_{r.m.s.} = \sqrt{\frac{h_1^2 + h_2^2 + h_3^2 + - h_n^2}{n}}$$

$$\text{or } h_{r.m.s.} = \left(\frac{1}{L}\int_0^L h^2 \, \mathrm{d}L\right)^{\frac{1}{2}}$$

9.43 Centre Line Average Method (R_a)

The R_a value is the standard adopted in Great Britain, and since 1955, the U.S.A. It is defined as the average height from a mean line of all ordinates of the surface, regardless of sign.

Thus, referring to Fig. 9.6,

$$R_a = \frac{h_1 + h_2 + h_3 + h_5 + h_6 \ldots + h_n}{n}$$

$$= \frac{\Sigma h}{n}$$

To determine an R_a value by the erection of ordinates would be a laborious process, and if an unfortunate ordinate spacing was chosen a significant point on the surface could be missed.

However, if an irregular area is divided by its length then the value obtained is the average height of the area. Such an area can be measured using a planimeter, thus considering an infinite number of ordinates, and every point on the surface is considered. Referring to Fig. 9.7 it is seen that

$$R_a = \frac{A_1 + A_2 + A_3 + \ldots + A_n}{L}$$

$$= \frac{\Sigma A}{L}$$

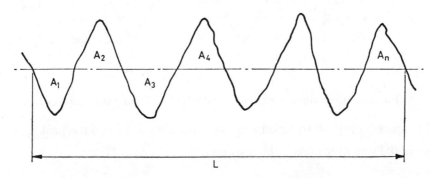

Fig. 9.7. Graphical representation of R_a.

The value thus obtained is the average height of the *trace*. To obtain the R_a value of the surface it is necessary to divide this value by the vertical magnification of the trace, and multiply by 10^6 to give the value in micro-inches.

$$\therefore R_a = \frac{\Sigma A}{L} \times \frac{10^6 \ \mu\text{in}}{\text{vertical magn.}}$$

where ΣA = sum of areas above and below the mean line in inches

$$L = \text{length of trace in inches}$$

or in metric units, $R_a = \dfrac{\Sigma A}{L} \times \dfrac{10^3 \ \mu m}{\text{vertical magn.}}$

where ΣA = sum of areas in mm²

L = length of trace in mm

Before making such a measurement it is necessary to position the mean line so that the areas above and below it are equal to within 5%. This can be done by:

(a) Estimating its position by eye.

(b) Measuring the total areas above and below the estimated line.

(c) Correcting the position an amount equal to the difference in areas divided by the trace length.

i.e. Correction $= \dfrac{\Sigma A \ (\text{above}) - \Sigma A \ (\text{below})}{L}$

This procedure can be illustrated by referring to a hypothetical surface having triangular irregularities as shown in Fig. 9.8.

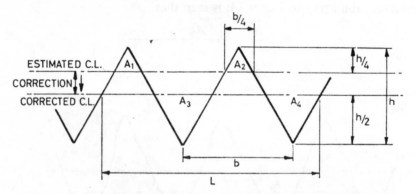

Fig. 9.8. Correction of estimated centre line for regular configuration.

Let an estimated centre line be drawn a distance $h/4$ from the peaks.

Then the sum of the areas above estimated line $= A_1 + A_2$

$$= \frac{2}{4} \cdot \frac{h}{4} \cdot \frac{b}{2} \cdot \frac{1}{2} = \frac{hb}{16}$$

And the sum of the areas below estimated line $= A_3 + A_4$

$$= \frac{2}{4} \cdot \frac{3h}{4} \cdot \frac{3b}{2} \cdot \frac{1}{2} + = \frac{9hb}{16}$$

$$\text{Correction} = \frac{\text{Area above} - \text{Area below}}{\text{Length}}$$

$$= \frac{\dfrac{hb}{16} - \dfrac{9hb}{16}}{2B} = -\frac{h}{4}$$

i.e. the correction distance for the trace centre line is $h/4$ downwards (as shown by the minus sign). This brings it to $\frac{h}{2}$ from the peaks, which would be correct.

This process is geometrically correct for a regular profile of this type, and is approximately correct for an irregular profile of the type encountered in a surface recording.

The technique described above is that required for the Tomlinson instrument. The Talysurf average meter incorporates an integrating device which does this work while the probe is being drawn over the surface.

9.5 THE INTERFERENCE MICROSCOPE

Reference to Chapter 2 shows that if an optical flat is placed at a small angle over a flat reflecting surface, and the system is viewed in a parallel beam of mono-chromatic light, interference fringes will be seen across the surface. The curvature

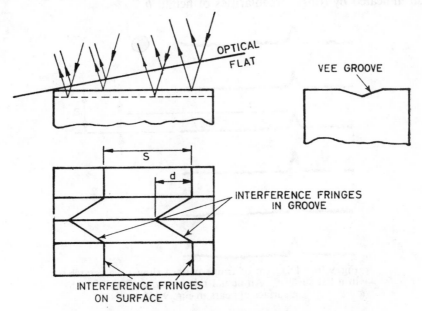

Fig. 9.9. Interference fringes across a flat surface having a shallow vee groove along its length.

of these fringes is a measure of the flatness of the surface and if the wavelength of the light used is known then the flatness can be determined to a high degree of accuracy.

Consider the effect of viewing a surface containing a 'vee' groove of depth 0·1 μm, under an optical flat in light of wavelength of 0.5 μm. The irregularity in the fringe pattern would indicate the depth of the groove as shown in Fig. 9.9.

The distance between the fringes represents change in distance between the surface and the optical flat of half a wavelength $\lambda/2$.

$$\therefore \text{ Depth of groove} = \frac{d}{s} \times \frac{\lambda}{2}$$

and in this case depth $= 0\cdot1 \ \mu\text{m}$

$$\lambda = 0\cdot5 \ \mu\text{m}$$

$$\therefore \ 0\cdot1 = \frac{d}{s} \times \frac{0\cdot5}{2}$$

$$\therefore \ \frac{d}{s} = \frac{0\cdot1}{0\cdot25} = 0\cdot4$$

It follows that the fringe irregularity, expressed as a fraction of the total fringe spacing, indicates the depth of surface irregularities in terms of the half wavelength of the light being used.

A surface appearing as in Fig. 9.10 has a general fringe spacing of a with a scratch indicated by fringe irregularities of height b.

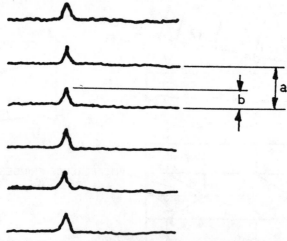

Fig. 9.10. Diagram of fringe pattern showing a scratch in a flat surface. An actual microinterferogram of such a surface appears in Fig. 9.11(a).

The depth of the scratches is therefore $\frac{b}{a} \times \frac{\lambda}{2} \ \mu\text{m}$ where λ is the wavelength of light used.

This work cannot be done with the normal optical flat set-up, but it requires a relatively steep angle to the optical flat, giving very fine fringes, and a microscope with high resolving power to enable them to be seen. Such a system enables interference systems to be observed on the surfaces of steel balls, the roundness of the fringes indicating the roundness of the balls, any surface irregularities

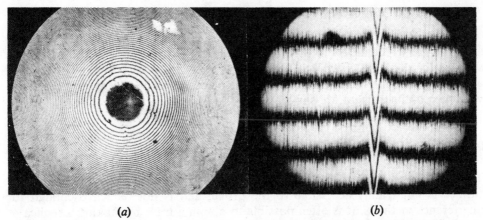

(a)

(b)

Fig. 9.11. Microinterferograms of (a) the surface of a steel ball; (b) a scratch on a flat surface approximately $\lambda/2$ or 0·25 μm deep.

(Courtesy of Hilger & Watts Ltd.)

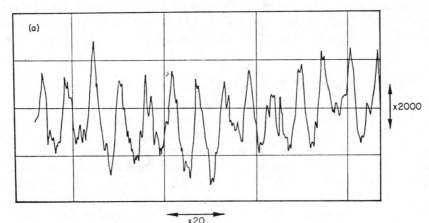

Fig. 9.12(a). Trace of surface. R_a reading = 3·65 μm.

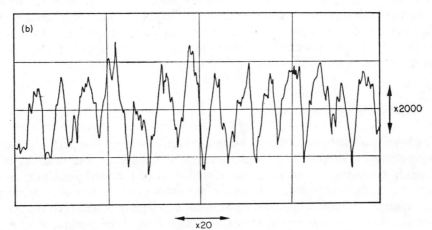

Fig. 9.12(b). Trace of replica of surface. R_a reading = 3·45 μm.

appearing as irregularities in the interference fringes. An interference microscope suitable for this type of work is made by Messrs. Hilger & Watts, incorporating a camera enabling permanent records to be made. Microinterferograms produced with such an instrument are shown in Fig. 9.11 (*a*) and (*b*).

9.6 REPLICA METHODS

There are cases where the surface texture of a component is required but the surface is not readily accessible to the probe, and is not reflective enough for interference methods. It is often possible to obtain a trace by making a replica of the surface. This was originally done by damping a piece of cellulose-acetate film in acetone to soften it, and pressing against the surface until it hardened. By this method a reproduction fidelity of about 80% could be obtained. Modern techniques use epoxy and other resins, and approach 100% fidelity. A kit of this type is available from Taylor–Hobson, and Figs. 9.12 (*a*) and (*b*) are traces of the same surface taken from the actual surface, and from a replica, respectively.

9.7 ASSESSMENT OF ROUNDNESS

Roundness is a geometric property of a cylindrical workpiece, errors in which are caused by errors in the geometry of the machine producing the part. Thus, deviations from a perfect circle are errors in macro-geometry rather than micro-geometry and it may well be asked why is the assessment of roundness associated with surface texture in this work. There are two reasons for this, the first being that in rotary bearings the two factors affecting the efficiency of the bearing are the surface texture and roundness of the mating parts. The second reason is that much of the fundamental work on the assessment of roundness was carried out by the Taylor-Hobson division of Rank Precision Industries Ltd. and the main amplifier used in their Talysurf instrument is used in the Talyrond which they developed for the assessment of roundness. For these two reasons surface texture and roundness assessment tend to be associated although they are different types of measurement requiring different techniques.

9.71 Roundness Testing Machines

If a dial gauge could be caused to move in a perfectly straight line path, the plunger bearing against a supposedly straight workpiece, changes in the readings of the dial would consist of the errors in the straightness of the workpiece plus a component due to the misalignment of the surface datum and the line of movement of the dial gauge. If the misalignment component can be eliminated then the remaining variations in readings will be errors in straightness of the workpiece.

Similarly, if a measuring instrument can be caused to move in a perfect circle, its detector or stylus bearing against a supposedly cylindrical workpiece, then the variations in readings will include deviations from roundness of the workpiece plus a component due to the misalignment of the axis of rotation of the measuring device and the axis of the workpiece.

It follows that a machine for the assessment of roundness must include the following features:

(a) A precision bearing, the axis of rotation of which gives an accurate datum for the measurement
(b) A means of mutually aligning the axes of the work and bearing
(c) A measuring device capable of amplifying errors in roundness
(d) A suitable recorder to give a visual indication of errors
(e) A means of making a quantitative assessment of errors.

These features are shown in the form of a block diagram in Fig. 9.13. Note that two alternative forms are available, one with a fixed table and rotating measuring head as shown, the other carrying the workpiece on an adjustable rotating table with a fixed measuring instrument. In either case the principles involved are the same.

9.711 Bearings for Roundness Testing Machines

The heart of any roundness testing machine is the bearing, for any errors in its manufacture which will cause a movement of the axis of rotation (run-out) will introduce errors in the final result which will be very difficult to isolate. The 'Talyrond', probably the best known of the machines available, utilizes precision needle roller bearings. Such bearings are extremely expensive to produce to the accuracy required and in recent years their precision has been matched by air bearings in which the moving surfaces are separated by films of air under pressure. These bearings are almost friction free and, if correctly designed, are extremely stiff, i.e. the deflection per unit load is very small. Such a bearing is used in the machine produced by Optical Measuring Tools Ltd. It was developed by the National Engineering Laboratory and is claimed to revolve on a constant axis to within 0·05 μm. It must be emphasized that if the accuracy of these bearings is to be maintained they should not be rotated under conditions of metal-to-metal contact; that is before air pressure is applied.

9.712 Alignment of Work and Rotational Axes

This is usually achieved by mounting the worktable on a pair of coordinate slideways as shown in Fig. 9.13. The adjustment of the table is by means of a pair of fine thread screws, the alignment being initially set by eye, then with the machine on low magnification and finally at the desired magnification.

191

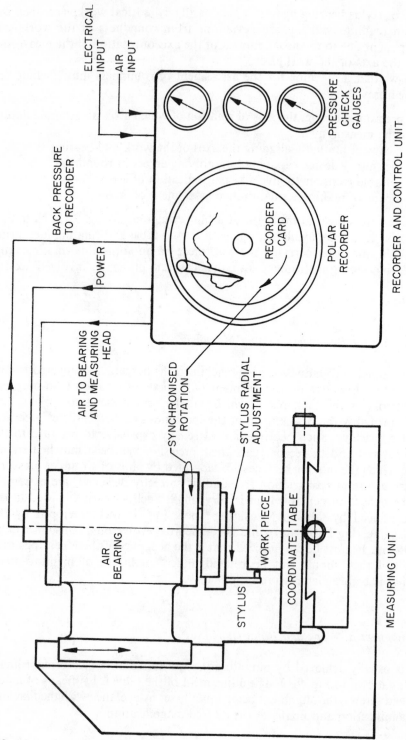

Fig. 9.13. Block diagram of a roundness testing machine.

9.713 Measuring Device

The measuring probe may be moving on the rotating head machines or fixed on the rotating work machine. In either case the measuring probe will be remote from the recording unit and will normally require flexible connections to it. Measuring methods lending themselves to such applications are electrical and pneumatic gauging.

The Talyrond machine uses electrical methods similar to those of the Talysurf (see 9.33) with their attendant advantages of easy change of amplification and an electrical output which can be used as the input to a special purpose computer which gives a direct reading of numerical assessment of non-roundness.

The O.M.T. machine uses an air-gauging system with two levels of pneumatic amplification which can be changed simply by adjusting a valve which changes the area of the control orifice. A second method of changing the amplification is by changing the length of the stylus arm. As the distance from the fulcrum to the measuring jet is constant a change in stylus arm length produces a change in amplification.

9.714 Recording Device

In all of these instruments the recorders used are polar type rather than the linear type. Consider a perfectly round component being checked on a machine using a linear recorder. If the axes of work and rotation are misaligned the trace produced will be a sine wave whose amplitude will be equal to the misalignment multiplied by the magnification being used. If the workpiece deviates from true roundness then these deviations will be superimposed upon the sine wave and the two components will be very difficult to separate. If, however, the trace is produced on a polar recorder whose rotation is synchronized with that of the bearing, then the deviations from roundness will be superimposed on a circle whose centre is offset from that of the recorder by an amount equal to the misalignment of the axes multiplied by the magnification. Thus the shape of the trace will be the same regardless of the magnitude of the misalignment, it will simply be in a different position on the recorder paper, and thus much easier to analyse.

The Talyrond recorder is an electrical device, similar in principle to that used for the Talysurf, as indeed is the amplifier unit. The O.M.T. recorder is basically a large, extremely sensitive Bourdon tube type pressure gauge whose pointer carries a tracer pen which moves over the recorder paper. Both recorders use discs of suitable paper which are caused to make one revolution as the measuring head makes one revolution, the motor drive being arranged to give just one complete revolution.

9.72   Assessment of Roundness Errors

It is important to note that it is the *errors* or deviations from roundness which are magnified and recorded. The mean diameter of the trace produced depends upon the initial setting of the machine and in any case it cannot exceed the size of the recorder paper. This can give a very misleading impression of the geometry of a workpiece, the effect being illustrated in Fig. 9.14 which shows traces produced from the same workpiece at different magnifications. The simplest way to make a numerical assessment of non-roundness is to place over the trace a transparency of concentric circles and find by trial and error the position giving the minimum difference in diameter between the two circles which just contain, and are contained by the trace. The radial distance between these circles, divided by the magnification, is then the error in roundness. This method is not very satisfactory being based upon trial and error. Neither is it systematic, one operator possibly selecting the suitable inner circle first and another the outer circle first. These methods could give different results.

B.S. 3730 suggests four methods of obtaining a numerical assessment of roundness from a trace, two being non-preferred and two preferred methods. They are all based on finding the centre of a particular circle and from that centre drawing the two circles which just contain and are contained by the trace. The deviation from roundness is the radial separation of the circles divided by the magnification. This would appear to be similar to the method suggested in the previous paragraph but it is the systematic method of finding the centre of the circles which is important.

9.721   *Plug Gauge Centre (PGC)*

This is the centre of the largest circle which is just contained by the trace. It is found by trial and error and will contact the trace at least at two points, possibly more. The outer circle is also drawn from this centre and will probably contact the trace at one point. The deviation from roundness is then the radial distance between the circles divided by the magnification.

9.722   *Ring Gauge Centre (RGC)*

The ring gauge centre is the centre of the smallest circle which will just contain the trace. Again it is found by trial and error and will contact the circle at a number of points on its periphery. The inner circle is also drawn from this centre and the deviation from roundness is found as for the PGC.

The above methods are both based upon trial and error and different operators could obtain different results. For this reason both are non-preferred.

194

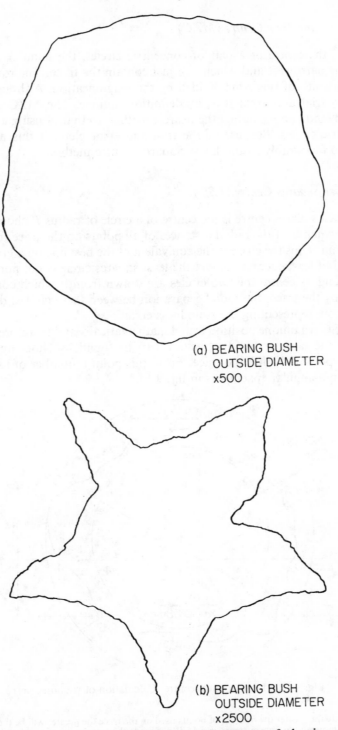

(a) BEARING BUSH
OUTSIDE DIAMETER
x500

(b) BEARING BUSH
OUTSIDE DIAMETER
x2500

Fig. 9.14.   Roundness traces from outside diameter of a bearing
bush at different magnifications.

9.723 Minimum Zone Centre (MZC)

This is the centre of a pair of concentric circles, the annular zone between which is the narrowest and which will just contain the trace. The roundness error is then the width of this zone divided by the magnification. Although this centre is found by trial and error it is, by definition, unique. The MZC is therefore a preferred method and is in fact the nearest method to that of using a transparency of concentric circles. Because of the trial and error element this method is not regarded so favourably as the Least Squares Centre method.

9.724 Least Squares Centre (LSC)

The least squares centre is the centre of a circle of radius R chosen so that the sum of the squares of the radial distances of all points on the trace from the circle is a minimum. Thus the circle is the equivalent of the best line on a graph drawn by the method of least squares. In fact the least squares circle is not normally drawn. Having found its centre the two circles are drawn from it which contain and are contained by the trace, the radial separation between these circles, divided by the magnification, representing the roundness error.

The LSC is a unique position found mathematically and therefore all operators should get the same result. Its position can be found by choosing as an initial datum any point O within the trace. From this point a number of radial lines are drawn at equi-angular spacing as in fig. 9.15.*

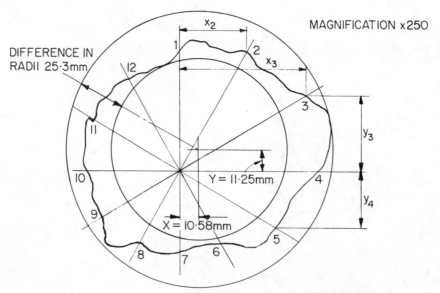

Fig. 9.15. Least squares centre. Calculation of roundness error.

* Obviously the greater the number of points used on the trace the greater will be the accuracy of the result but in practice little improvement is achieved if more than twelve divisions are used.

From the intersection of these lines with the trace the horizontal coordinates x_1, x_2, x_3, x_4, etc., and the vertical coordinates y_1, y_2, y_3, y_4, etc., are measured relative to point O. The coordinates X and Y of the least squares centre from this point are then found from:

$$X = \frac{2\Sigma x}{n} \text{ and } Y = \frac{2\Sigma y}{n}$$

where Σx = sum of all values of x
Σy = sum of all values of y
n = number of positions considered

This calculation is best set out in tabular form, the calculation below referring to Fig. 9.15.

Position	x mm	y mm
1	0	63·0
2	36·5	62·5
3	69·5	40·5
4	75·0	0
5	52·0	− 30·0
6	22·0	− 38·5
7	0	− 43·0
8	− 22·5	− 38·5
9	− 43·0	− 24·5
10	− 51·0	0
11	− 47·0	27·0
12	− 28·0	49·0
Totals	63·5	67·5

$$X = \frac{2\Sigma x}{12} = \frac{127}{12} = 10 \cdot 58 \text{ mm}$$

$$Y = \frac{2\Sigma y}{12} = \frac{135}{12} = 11 \cdot 25 \text{ mm}$$

Difference in radii = 25·3 mm

$$\text{Roundness error} = \frac{\text{Difference in radii}}{\text{Magnification}} \times 1000 \ \mu\text{m}$$

$$= \frac{25 \cdot 3}{250} \times 1000 \ \mu\text{m}$$

$$= 101 \cdot 2 \ \mu\text{m}$$

9.73 Concentricity

Apart from roundness other forms of error in surfaces of revolution can be checked using roundness testing equipment. If a component has been set up for testing the roundness of its bore then by changing the measuring head to read on an outside diameter the two traces will be superimposed. The separation of the centres of the traces divided by the magnification will be equal to the eccentricity. Two such traces are shown in Fig. 9.16. It should be noted that this should only be done if the instrument head is not raised or lowered. Such a movement may cause a realignment of the measuring head and work, which can induce an error.

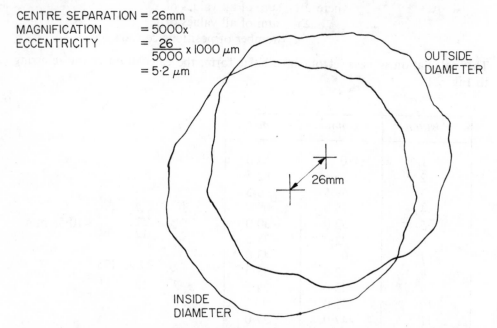

CENTRE SEPARATION = 26mm
MAGNIFICATION = 5000x
ECCENTRICITY = $\dfrac{26}{5000}$ x 1000 μm
= 5·2 μm

OUTSIDE DIAMETER

26mm

INSIDE DIAMETER

Fig. 9.16. Traces of O/D and I/D of bearing bush superimposed from same setting showing eccentricity.

CHAPTER 10

Statistical Quality Control

10.1 INTRODUCTION

THE previous chapters in this book tend to emphasize the fact that although accuracy is of great importance to engineers, absolute accuracy is unattainable. Even gauge blocks of the highest order of accuracy have manufacturing tolerances which are as small as a few millionths of an inch. These tolerances, whose magnitude depends on the function of a component, are necessary to allow for the inherent variability of the production process.

Consider a simple operation such as a capstan lathe set to part off lengths of bar stock. Within limits, the lengths parted off will be the same, but variations around the length at which the machine is set will occur. These variations may be due to a number of causes any one of which will be of negligible effect, but if all causes tend to produce a size increase then a piece of maximum length will be produced. Similarly, if all causes tend towards a decrease a piece of minimum length will occur, and if all causes work against each other their effects will be cancelled and a piece close to the set size will be cut off.

These random variations are due to non-assignable causes. It is the function of a system of process control to distinguish between these and other, assignable, causes such as the machine setting changing, a grinding wheel wearing or a machine failure occurring. These assignable faults must be detected and corrected before defective pieces are produced.

Apart from component production, faults can occur during assembly, and a large manufacturer buying in parts or sub-assemblies must control the quality of the goods he is buying. This also can be achieved by the intelligent application of statistical methods.

This chapter will therefore consider the ways in which statistics can be used to perform these functions in an economical manner. However, it should be borne in mind that there is no substitute for 100% inspection, and even then, human nature being what it is, faulty parts may be accepted by the inspector.

10.2 PROCESS VARIABILITY

It has been pointed out in section 10.1 that all manufacturing processes, and indeed many natural ones, are subject to random variations. This may be shown by a simple experiment which consists simply of measuring about 200 parts and recording their sizes. If simple parts such as dowel pins, produced by centreless grinding, are used their diameters should be recorded to the nearest 0·001 mm. This does not mean that their diameters should simply be noted. It would be a laborious operation and would yield little useful information.

The most convenient way to record this information is to set up a 'tally chart', as in Fig. 10.1. The diameters of about ten dowel pins should be measured with a micrometer and their mean size found. A comparator of suitable magnification can now be set to this size and as each part is passed under the measuring head its deviation + or − from this mean size is noted by a mark on the chart, measurements being made to the nearest 0·001 mm unit.

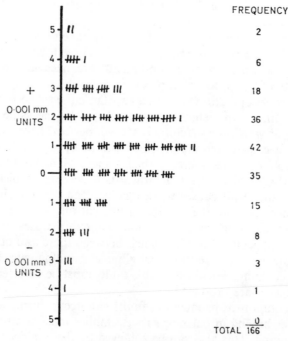

Fig. 10.1.   Tally chart.

This tally chart itself shows some of the characteristics of the process variability but they are better shown if plotted to scale as a graph of frequency against size. Such a graph, known as a frequency polygon, is shown in Fig. 10.2.

This frequency polygon is a typical representation of a manufacturing process where size is subject to random variation. It can usually be shown that under these circumstances the frequency polygon is very similar to the normal or Gaussian curve shown in Fig. 10.3.

The two axes of this curve represent the work size (horizontal) and the number of times each work size occurs, or frequency (vertical). The total spread

200

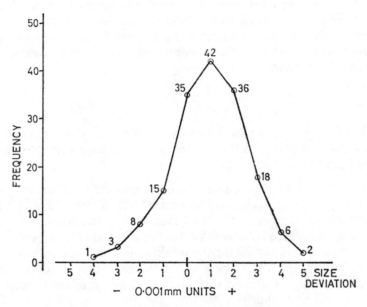

Fig. 10.2. Frequency polygon of results in Fig. 10.1.

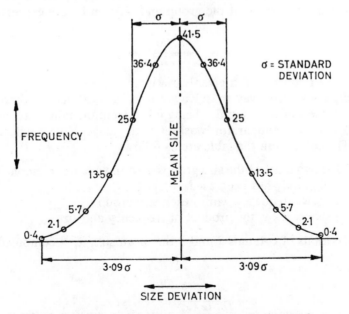

Fig. 10.3. Normal distribution curve fitted to results from
Fig. 10.1.

of the curve are the limits between which work will normally be produced. How-ever, a batch of parts may not give components of limiting size range. A better guide is the standard deviation which is the distance from the vertical axis of the normal curve to the point where its curvature changes from convex to concave.

The law of this curve is

$$y = \frac{1}{\sigma\sqrt{(2\pi)}} e^{-\frac{(x-\bar{x})^2}{2\sigma^2}}$$

in which π and e are the mathematical constants; $\bar{x}$ is the mean size; x is the size of the individual part; y is the frequency with which a size x occurs; and σ is the standard deviation.

It is the properties of this curve which are important in quality control, the most important property being the fact that 99·8% of the area under the curve, which represents to some scale the total number of parts considered, lies within the range of $\pm 3\cdot 09\sigma$ from the mean, i.e. all but 1 in 1000 at each tail.

Thus, if a sample of say 2000 parts is taken, and the standard deviation calcu-lated, it can be assumed that as long as the machine stays in control, no change taking place in setting or process variability, all parts produced except 2, one at each extreme, will lie within $\pm 3\cdot 09\sigma$ from the mean size.

10.21   Calculation of Standard Deviation

The standard deviation may be defined as the root mean square of the individual deviation from the mean size of the group and is given by the expression,

$$\sigma = \sqrt{\frac{\Sigma F (x-\bar{x})^2}{\Sigma F}}$$

in which F is the frequency of a given variation.*

The most convenient way to calculate σ is by a tabular method as shown in section 10.4, the values of x and F being taken from the tally chart in Fig. 10.1.

Assume that the comparator was set with gauge blocks whose size was 6·005 mm. The columns in the table are as follows:

Column 1 shows a list of the size groups used in which, for simplicity, the last
digit only has been used.
Column 2 shows frequency within each size group.
Column 3 shows $F \times x$, the product of frequency and size.

From the totals of columns 2 and 3 the mean size, $\bar{x}$, is computed.

$$\bar{x} = \frac{\Sigma (Fx)}{\Sigma F}$$

---

* A better value is given by $\sigma = \sqrt{\frac{\Sigma F (x-\bar{x})^2}{(n-1)}}$ where n is the number of parts in the group.

1		2	3	4	5	6
Size x	*Code* x	*Frequency* F	$F \times x$	$x - \bar{x}$	$(x - \bar{x})^2$	$F(x - \bar{x})^2$
6·001	1	1	1	−5	25	25
6·002	2	3	6	−4	16	48
6·003	3	8	24	−3	9	72
6·004	4	15	60	−2	4	60
6·005	5	35	175	−1	1	35
6·006	6	42	252	0	0	0
6·007	7	36	252	+1	1	36
6·008	8	18	144	+2	4	72
6·009	9	6	54	+3	9	54
6·010	10	2	20	+4	16	32
Totals:		166	988			434

Fig. 10.4. Tabular calculation of standard deviation based on frequency distribution in Fig. 10.1.

$$\text{Mean size } \bar{x} = \frac{\Sigma F x}{\Sigma F}$$

$$= \frac{988}{166} = 5 \cdot 95 \text{ (0·001 mm units)}$$

$$\therefore \bar{x} = 0 \cdot 006 \text{ mm}$$

$$\sigma = \sqrt{\frac{\Sigma F (x - \bar{x})^2}{\Sigma F}}$$

$$\sigma = \sqrt{\frac{434}{166}} = 1 \cdot 6_{14} \text{ (0·001 mm units)}$$

$$\therefore \sigma = 0 \cdot 0016 \text{ mm}$$

Column 4 shows the deviation of each size group from the mean, i.e. $(x - \bar{x})$.

Column 5 shows the square of the figures in column 4, i.e. $(x - \bar{x})^2$.

Column 6 shows the product of frequency (column 2) and deviation squared (column 5), i.e. $F(x - \bar{x})^2$.

Then σ is obtained by dividing the total of column 2 into the total of column 6 and finding the square root.

In this process the standard deviation = 0·0016 mm
then process variability = $\pm 3 \cdot 09 \sigma = \pm (3 \cdot 09 \times 0 \cdot 0016 \text{ mm})$
$= \pm 0 \cdot 0049$ mm (to nearest 0·0001 mm)

203

Therefore, at one machine setting, the size produced is unlikely to vary from the net size by more than ±0·0049 mm. If a work tolerance of ±0·005 mm was given then there would be a probability of some scrap being produced as soon as the setting changed, i.e. if the grinding wheel or tool started to wear.

10.22 Standard Deviation of the Average Size of Sub-groups

If a large batch of parts was taken and split at random into groups of, say, 5 parts, then the average size of the sub-groups could be measured. A little thought will show that these average sizes would be grouped more closely around the mean size than would the sizes of the individual items. Typical frequency distributions for the average sizes and the individuals are shown in Fig. 10.5.

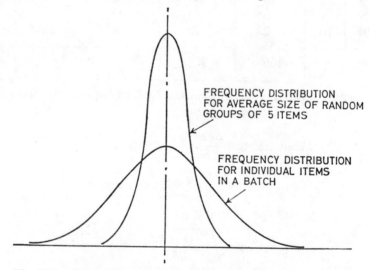

FREQUENCY DISTRIBUTION FOR AVERAGE SIZE OF RANDOM GROUPS OF 5 ITEMS

FREQUENCY DISTRIBUTION FOR INDIVIDUAL ITEMS IN A BATCH

Fig. 10.5. Comparison of distribution curves for individual items and group averages based on results from Fig. 10.1.

If σ_g is standard deviation for group means and σ_b is standard deviation for individuals in the batch, then it can be shown that:

$$\sigma_g = \frac{\sigma_b}{\sqrt{n}} \text{ where } n \text{ is the number of items in the group}$$

This point is most important in the application of these principles to control charts.

10.3 CONTROL CHARTS

If a machine is set to produce work to a given size two changes in product quality can occur. These are:

(*a*) The mean size of the work produced may change, i.e. the setting has moved, e.g. tool wear has occurred.

(*b*) The process variability may change, i.e. the size range of work produced has changed. This is usually due to something more fundamental in the machine.

In either case scrap will be produced unless corrective action is taken.

The effects of these two changes are shown in Figs. 10.6 and 10.7 in which, initially, the work was within specified limits.

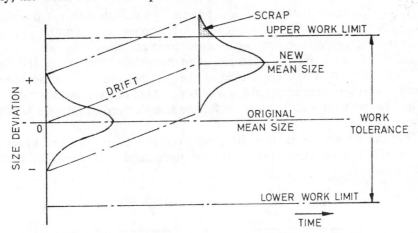

Fig. 10.6. Effect of mean size drifting during a process.

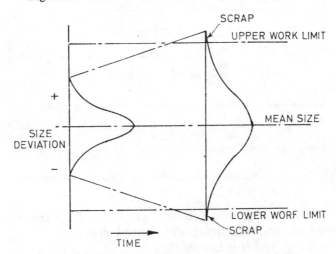

Fig. 10.7. Effect of a change in process variability.

If these effects occur either singly or together, the final outcome will be the same—defective work which may or may not be reclaimable and which, if mixed with good work, will require 100% inspection to detect.

Thus control charts are required for controlling the setting size—*average charts*; and for controlling the process variability—*range charts*.

10.31   Control Charts for Average

The simplest form of control chart would be one on which the upper and lower work limits are shown and on which the size of every workpiece, as it is produced, is plotted.

This would show both range and setting, but unfortunately it is simply an expensive method of 100% inspection, and therefore not an acceptable alternative.

Consider an inspector coming to a machine at stated intervals, taking say five successive parts known as a sample, computing their average size and plotting this on such a chart. The average sizes of the groups would vary about the mean size and when the setting changed a dimensional trend in a certain direction would appear. The problem is to decide how far the sample average size can be allowed to drift.

This can be seen from Fig. 10.8, which shows the upper and lower work limits, a distribution curve for individual items and a distribution curve for sample averages.

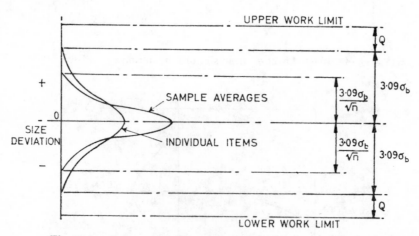

Fig. 10.8.   Construction of a control chart for sample averages.

It can be seen that the process can be allowed to drift an amount $\pm Q$ before there is a significant chance that scrap will be produced.

Further from Fig. 10.5 it is known that

$$\sigma_g = \frac{\sigma_b}{\sqrt{n}}$$

$$\therefore \ \pm 3 \cdot 09 \sigma_g = \pm \frac{3 \cdot 09 \sigma_b}{\sqrt{n}} = \text{total spread of sample averages curve}$$

206

Therefore if the process remains in control (statistically) the points plotted for the sample averages should always be within $\pm\dfrac{3\cdot09\sigma_b}{\sqrt{n}}$ of the required mean size.

However, the process can drift from this mean setting an amount $\pm Q$ before scrap is produced, Q depending on the difference between the process variability and the work tolerance.

Thus the maximum distance a sample average should be away from the mean size is $\pm\left(Q+\dfrac{3\cdot09\sigma_b}{\sqrt{n}}\right)$. As soon as a point occurs outside these limits it should be investigated, as the deviation is greater than that occurring due to random, or non-assignable, causes.

The control chart for averages is drawn as shown in Fig. 10.9.

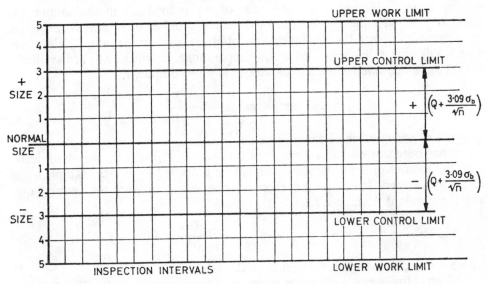

Fig. 10.9. Control chart for average size of samples —$\bar{x}$ chart.

Immediately the points plotted show a definite drift to one side or the other of the mean size, the process should be treated as suspect. Before the trend reaches the control limits it can be stopped and the process reset. Note that in a process where the drift is usually in one direction, as on a grinding operation where wheel wear occurs, the resetting can be purposely towards the other control limit, thus ensuring larger runs between resetting.

10.311   *Simplified Method of Setting Control Limits*

Fixing the position of the control limits relative to the mean size would appear to be a laborious process if the method previously described involving a calculation of the standard deviation, is used. However, it should be realized that this is intended to illustrate the fundamentals of the system. The standard deviation is in fact a measure of the range of work sizes which will be produced in a given process at one setting. Thus there should be some relationship between standard deviation and the size range in the samples inspected and plotted on the chart. This is so, and it greatly simplifies the work in setting up a control chart.

When a chart is in operation the inspector taking the sample and plotting its average size, should carry out his calculation on a properly designed data sheet, not an odd scrap of paper. If this is done then the same data sheet can be used in setting up the chart.

To set up the chart the inspection frequency and the sample size should be decided upon, and this form of inspection carried out, *in association with the existing method.* Having noted the sizes of the individuals in the sample the inspector computes the average size of the sample and also the *sample range,* i.e. the difference between the maximum and minimum sizes in the sample. Thus over a period of time a number of values of sample range, denoted by w, are available.

The *average value of the sample ranges* is now computed.

$$\text{Mean sample range} = \bar{w} = \frac{\Sigma w}{n_s}$$

where n_s is the number of samples taken

It has been shown that a measure of process variability is the standard deviation, but to control the process variability it is proposed to plot the sample range on a control chart. It follows that there is a connection between the sample range and the standard deviation, i.e.

$$\sigma = C\bar{w} \qquad \text{and} \qquad \therefore \quad \frac{3 \cdot 09\sigma}{\sqrt{n}} = A\bar{w}$$

where A is a constant which depends on the sample size n.

Control limits may now be set at the specification mean $\pm A\bar{w}$. This takes no account of the relationship between the work tolerance and the process variability and makes no provision for the process to be allowed to drift the amount Q. To determine whether the process may be allowed to drift before corrective action is taken, the *Relative Precision Index* (R.P.I.) is used. This relates process variability to work tolerance by

$$\text{R.P.I.} = \frac{\text{work tolerance}}{\bar{w}}$$

Having calculated the R.P.I., Table 4 on page 211* is examined against the

* Taken from B.S. 2564.

appropriate sample size to see whether the process is of low, medium or high relative precision.

If low, the process is unsatisfactory and likely to produce a proportion of defective work, i.e. Q is negative.

If medium, the process is satisfactory and, if well controlled, will produce no defectives. In this case Q is zero and the process will require frequent adjustment to maintain satisfactory control. The control and warning limits can be set by multiplying $\bar{w}$ by the appropriate values of the constant A' in Table 2 on page 202.

If the process is one of high relative precision a drift can be allowed to take place before the setting is adjusted, i.e. Q is positive. In this case the control chart limits are found by multiplying $\bar{w}$ by the appropriate constants in Table 5. In this case, to allow for the process to drift, the limits *are set in from the work limits by an amount $A''\bar{w}$*.

Thus in all cases the control chart limits are set by multiplying the mean sample range $\bar{w}$ by constants provided in tables in B.S. 2564.

10.32   Control Charts for Range

It was shown in section 10.3 that, as well as a control chart for averages, a range chart to control the process variability is required. The control limits for the range chart are set in a similar manner to those for the average chart, i.e. constants have been arrived at which, when multiplied by the mean sample range, $\bar{w}$, give the positions of the upper and lower limits for the range chart. These constants are shown in Table 3 of B.S. 2564 overleaf.

(*Text cont. on page 212*).

Extracts from B.S. 2564: 1965 are reproduced by permission of the British Standards Institution, 2 Park St., London W.1, from whom copies of the complete standard may be obtained. These tables are numbered as in the original.

TABLE 2

Control Chart Limits for Sample Average ($\bar{x}$)

To obtain limits multiply $\bar{w}$ by the appropriate value of $A'_{0.025}$ and $A'_{0.001}$ then add to and subtract from the gross average value or agreed objective ($\bar{X}$).

Sample size n	For inner limits $A'_{0.025}$	For outer limits $A'_{0.001}$
2	1·23	1·94
3	0·67	1·05
4	0·48	0·75
5	0·38	0·59
6	0·32	0·50

TABLE 3

Control Chart Limits for Range (w)

To obtain limits multiply $\bar{w}$ by the appropriate value of D'. To estimate standard deviation, σ, divide $\bar{w}$ by the appropriate value of d_n.

Sample size	For lower limits*		For upper limits		For standard deviation
n	$D'_{0.001}$	$D'_{0.025}$	$D'_{0.975}$	$D'_{0.999}$	d_n
2	0·00	0·04	2·81	4·12	1·13
3	0·04	0·18	2·17	2·99	1·69
4	0·10	0·29	1·93	2·58	2·06
5	0·16	0·37	1·81	2·36	2·33
6	0·21	0·42	1·72	2·22	2·53

* The lower limits are not generally used.

TABLE 4

Classification of Process Variability Relative to Specification Tolerance

Relative precision index. (R.P.I.) = specification tolerance/average range.

Note: This table should *not* be used if the range is out of control.

Class	Low relative precision	Medium relative precision	High relative precision
Sample size n	R.P.I.	R.P.I.	R.P.I.
2	less than 6·0	6·0 to 7·0	greater than 7·0
3	less than 4·0	4·0 to 5·0	greater than 5·0
4	less than 3·0	3·0 to 4·0	greater than 4·0
5 and 6	less than 2·5	2·5 to 3·5	greater than 3·5
State of production.	Unsatisfactory.* Rejections inevitable.	Satisfactory, if averages are within control limits.	Satisfactory, if averages are within modified limits.

* Not necessarily, if the specification permits a small proportion of the product to be outside the limit. In such cases the limiting values for low relative precision can be 0·8 of those given above.

TABLE 5

Modified Control Chart Limits for Sample Average ($\bar{x}$)

High precision class

To obtain the limits, multiply $\bar{w}$ by the appropriate values of $A''_{0.025}$ and $A''_{0.001}$, then add to the lower drawing limit and subtract from the upper drawing limit.

Number in sample n	For inner modified limits $A''_{0.025}$	For outer modified limits $A''_{0.001}$	For alternative* modified limits	
			Inner $A''_{0.025}$	Outer $A''_{0.001}$
2	1·51 (0·83)†	0·80 (0·12)†	2·32	1·61
3	1·16 (0·71)	0·77 (0·32)	1·70	1·31
4	1·02 (0·65)	0·75 (0·38)	1·46	1·19
5	0·95 (0·62)	0·73 (0·41)	1·34	1·12
6	0·90 (0·60)	0·71 (0·42)	1·26	1·08

When the inner and outer limits are close together, one of them can be omitted. (See also Part One, Section B4 *c*, page 24.)

* To provide higher assurance against manufacturing rejects.

† The alternative figures in brackets may be used when the bulk is permitted to contain a small proportion of rejects and the R.P.I. exceeds the value given in Column 2, Table 4.

A lower control limit for range may seem an anomaly, and in many cases it is not used. However, it should be remembered that statistically the range will be within the control limits and if it goes higher or lower it is due to an assignable cause. If a point on a range chart falls outside the upper control limit, the process should be stopped and investigated. If a point falls outside the lower control limit it means that for some reason the process variability has decreased and the process has improved. If the cause can be found it may be possible to incorporate the improvement in future similar operations and the overall quality improved.

10.33   Summary of Procedure in Setting up a Control Chart

This summary shows, step by step, the setting up of a control chart, and should be read with reference to Fig. 10.10 and Fig. 10.11 on facing pages 206 and 207. The data used is that from the tally chart in Fig. 10.1 and the data sheet, Fig. 10.10, may be used for both setting up and running a system of control charts.

It is recommended that the data sheet be printed on the back of the control chart, which should be of squared paper. They can then be kept in a transparent envelope at the machine in front of the operator, and filed when completed.

Step 1. Decide on the sample size *n* and the frequency of inspection. These should be chosen to give a total of 10–20% of total production. A good sample size is 7, but 5 is more convenient for computing averages.

Step 2. Take samples at the decided frequency and record their sizes on the data sheet.

Step 3. For each sample calculate sample average and sample range *w*.

Step 4. When enough parts (80–100) have been inspected and recorded calculate the mean sample range $\bar{w}$.

Step 5. Calculate Relative Precision Index.

$$\text{R.P.I.} = \frac{\text{work tolerance}}{\bar{w}}$$

Step 6. Find limits for $\bar{x}$ chart using Table 2 or Table 5 (B.S. 2564), depending on R.P.I.

Step 7. Set control limits for range chart using constants from Table 3.

10.4   CONTROL CHARTS FOR NUMBER DEFECTIVE

There are many cases in industry where the use of control charts is desirable, but is not possible in the form shown in section 10.3. Such cases are those where control of a manufacturing process by controlling a few dimensions is not convenient. Consider a production line producing an assembly which is to be checked by the customer by a sampling system which is based on the process producing an average of 2% defectives. It is desirable to ensure that on average the process is not producing an excess of 2% defectives. Other cases are those where completeness of the product rather than dimensional accuracy is required, e.g. the completeness of a plastic moulding or diecasting.

In instances such as these, a control chart for the number in a sample which are defective is desirable. These are number defective control charts in which the control limit is set at a value, for a given process average percent defective, above which the number defective in the sample will rise due only to an assignable cause. Variations in number defective in the sample below this level are to be expected and may be considered due to non-assignable causes.

If a process is running at an average of $G\%$ good parts and $B\%$, i.e. $(1 - G\%)$ defectives, and samples of n parts are taken, the number of defectives in the sample will vary around $B\%$ of n, which will be the 'expected' number of defectives in the sample. The distribution of the variation of the number of defectives is described very closely by the Poisson Probability Distribution, which is discussed in section 10.62.

Let $x =$ expected number of defectives in the sample

$= B\%$ of n where $n =$ sample size

Each term of the expansion of $e^{-x}e^{x}$ is the probability of 0 defectives, 1 defective, 2 defectives, 3 defectives, etc., appearing in the sample. Consider a process running at 2·5% defectives, a sample size of 150 being used.

$$x = 2\cdot5\% \text{ of } 150 = 3\cdot75$$

and $e^{-x} = 0\cdot0235$ (from tables of e^{-x})

$$\text{Now } e^{-x}e^{x} = e^{-x} + xe^{-x} + \frac{x^{2}e^{-x}}{2!} + \frac{x^{3}e^{-x}}{3!} + \frac{x^{4}e^{-x}}{4!} + \frac{x^{5}e^{-x}}{5!} + \cdots$$

$$= 0\cdot0235 + 0\cdot088 + 0\cdot163 + 0\cdot206 + 0\cdot194 + 0\cdot145 + 0\cdot091 +$$
$$0\cdot049 + 0\cdot028 + 0\cdot0095 + 0\cdot0035$$

To set the control limits on this type of chart it is necessary to determine the probability with which '*n*' *or more* defectives will occur in sample, for different values of '*n*'. The control limits are then usually set at that value of '*n*' *or more* having a probability of 0·05 for the Warning Limits and that value of '*n*' *or more* having a probability of 0·005 for the Action Limits. These are often known as the 1/20th and 1/200th limits respectively and can be found from tables published in B.S. 2564. The method detailed below, however, enables the limits to be determined to any probability level required.

From the foregoing figures obtained from the expansion of $e^{-x}e^{x}$ it can be seen that, if the expected number of defectives in a sample is 3·75 then:

Probability of 0 defectives in a sample $= 0\cdot0235$

Probability of *1 or more* defectives in a sample $= 1 - 0\cdot0235 = 0\cdot9765$

Similarly:

CONTROL CHART DATA SHEET | **G & S MANUFACTURING CO. LTD.** | **CONTROLLED** | **DATA SHEET NUMBER** 535/82

PART NAME _Dowel_ — INSPECTOR _J.C. Bloggs_ — DIMENSION _6·00mm +0·01mm +0·00mm DIA._ — SAMPLE SIZE _5_

PART Nº _XTH-535_ — OPERATOR _C.J. Jones_ — INSPECTION INT. 30 min.

MACHINE _Class Grinder_ — M/C Nº _182 DJH_

SAMPLE NUMBER

ITEM Nº	1	2	3	4	5	6	7	8	9	10	11	12	13	14	15	16	17	18	19	20	21	22	23	24	25	26	27	28	29	30	31
1	+1	+2	+4	-4	-3	-2	-1	+1	+2	-1	-2	+2	+1	0	+5	0	+1	+2	-2	+2	+3	-2	0	0	+3	+2	+1	+2	-1	+2	+2
2	-2	0	+2	0	+1	0	+1	+3	-1	+3	-1	+2	+2	0	+2	+3	+2	+2	+2	+1	0	+3	0	+2	0	+2	+5	+1	0	-1	+1
3	+3	+2	0	-1	+1	+1	0	0	-2	0	-1	0	+2	-2	0	-1	0	-1	+2	0	+2	+2	+1	+4	+2	+1	+2	+2	-1	+2	+3
4	+1	-3	-1	+1	0	+3	0	-2	+2	+1	0	+1	+1	+4	+1	+2	+2	0	+2	+1	-1	-1	+1	+2	+1	+3	+3	0	+3	+1	+2
5	0	+4	+1	+1	-1	+6	-3	+1	+1	+1	+3	0	0	0	+1	+2	0	0	+3	+1	+3	0	+2	+1	0	0	+1	+2	+2	+3	0
6																															
7																															
8																															
SAMPLE MEAN x̄	+0·2	+1	+0·4	-0·6	-0·4	+1·4	-0·6	+0·6	+0·4	+0·8	-0·2	+1	+1·2	+0·4	+1·8	+1·2	+1	+0·6	+1	+1	+1·4	+0·4	+0·8	+1·4	+1·2	+1·6	+1·6	+1·4	+1	+1·4	+1·6
SAMPLE RANGE w	6	7	6	6	4	7	4	5	4	4	5	2	2	6	6	4	2	3	5	3	3	5	2	4	3	3	2	2	3	4	3

CALCULATIONS

TOTAL OF w = 118

MEAN SAMPLE RANGE

$$\bar{w} = \frac{\text{TOTAL OF } w}{\text{Nº OF SAMPLES}}$$

$$= \frac{118}{31}$$

$$= 3·9 \;(0·0039mm)$$

RELATIVE PRECISION INDEX

$$\text{R.P.I.} = \frac{\text{WORK TOLERANCE}}{\bar{w}}$$

$$= \frac{0·010}{0·0039}$$

$$= 2·57$$

PROCESS IS OF _MEDIUM_ RELATIVE PRECISION

AVERAGE CHART

CONTROL LIMITS
= SPEC. MEAN $\pm (A'_{0·001}\, \bar{w})$
= $6·005 \pm (0·59 \times 0·0039)$
= $6·005 \pm 0·0023mm$

WARNING LIMITS
= $6·005 \pm (0·38 \times 0·0039)$
= $6·005 \pm 0·0015mm$

RANGE CHART

UPPER LIMITS

CONTROL = $2·36 \times 0·0039$
= $0·009mm$

WARNING = $1·81 \times 0·0039$
= $0·007mm$

LOWER LIMITS NOT USED IN THIS CASE

Fig. 10.10. Control chart data sheet.

This sheet is based on the data in the tally chart (Fig. 10.1) and should be examined in conjunction with the control chart (Fig. 10.11).

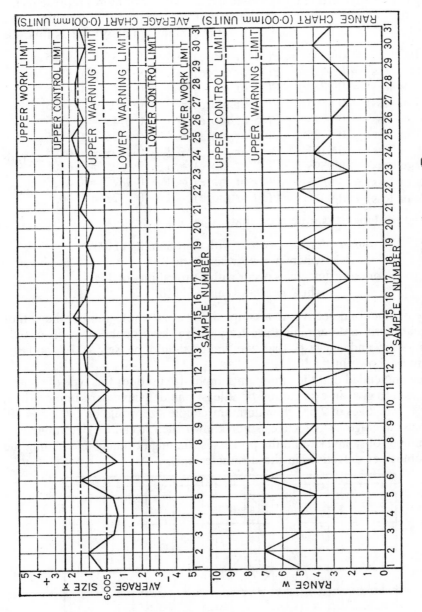

Fig. 10.11. Control charts for range (w) and average ($\bar{x}$).

These charts would normally be on the back of their data sheet (Fig. 10.10). Note the indication of drift on the $\bar{x}$ chart towards the U.C.L. Resetting will shortly be necessary.

Probability of *1 or less* defectives = 0·0235 + 0·088 = 0·1115
Probability of *2 or more* defectives in a sample = 1 − 0·1115 = 0·8885

Continuing in this way, letting P = probability, so that P(2 or more) represents the probability of 2 or more defectives in the sample, the probability values for a series of values of '*n*' or more defectives in the sample can be set out as follows:

P(0) = 0·0235	P(1 or more) = 1 − 0·0235 = 0·9765	
P(1 or less) = 0·1115	P(2 or more) = 1 − 0·1115 = 0·8885	
P(2 or less) = 0·2745	P(3 or more) = 1 − 0·2745 = 0·7255	
P(3 or less) = 0·4805	P(4 or more) = 1 − 0·4805 = 0·5195	
P(4 or less) = 0·6745	P(5 or more) = 1 − 0·6745 = 0·3255	
P(5 or less) = 0·8195	P(6 or more) = 1 − 0·8195 = 0·1805	
P(6 or less) = 0·9105	P(7 or more) = 1 − 0·9105 = 0·0895	
P(7 or less) = 0·9595	P(8 or more) = 1 − 0·9595 = 0·0405 ← 0·05	
P(8 or less) = 0·9875	P(9 or more) = 1 − 0·9875 = 0·0125	
P(9 or less) = 0·997	P(10 or more) = 1 − 0·997 = 0·003 ← 0·005	

From the above table it can be seen that:

'*n*' *or more* defectives occur with a probability of 0·05
when '*n*' is between 7 and 8
'*n*' *or more* defectives occur with a probability of 0·005
when '*n*' is between 9 and 10

As the number of defectives in a sample is always a whole number the author sees no great merit in determining precisely where the probability levels of 0·05 and 0·005 occur and suggests that the limits should be set at 7·5 and 9·5 respectively as shown in fig 10.12.

It is recommended that the minimum sample size used for this type of control chart is chosen to give an expectation of at least one defective per sample. Thus, with a PA% defectives of 2·5% the minimum sample size would be 100/2·5 = 40. If the process average % defective is not known then it can be determined by taking a number of samples when the process is running normally and finding the average number of defectives in these samples.

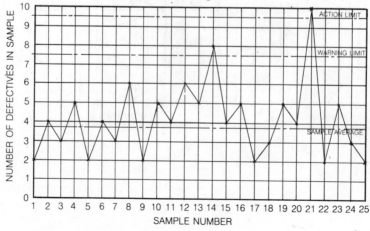

Fig. 10.12 Control chart for number defective.

10.41 Compressed Limit Gauges

As described above, control charts for number defective are suitable for final inspection on assembly lines and possibly plastic moulding and diecasting operations etc. In the case of machining processes the PA% defectives will be normally so low that the sample size will be excessively large and the technique impractical.

In such cases this type of chart can be used, the work being inspected using limit gauges set well inside the work limits so that they will always tend to reject a small proportion of work although it is not defective. This artificially high % rejects enables the chart to show when a significant change takes place in the process before it starts producing defective parts.

Parts rejected by the compressed limit gauges. A more detailed description of the use and design of compressed limit gauges is available in B.S. 2564.

10.5 SAMPLING INSPECTION OF INCOMING GOODS

Most industrial organizations buy in a considerable proportion of components and sub-assemblies from specialist manufacturers. In order to protect their own reputation, and that of their suppliers, it is necessary to control the quality of these goods. Quality assessment of large batches of parts can be carried out by three methods:

(a) *Spot checking*. This consists of inspecting a small sample here and there and hoping that the results are a reflection of the quality of the batch. The method is cheap but risky and furthermore, the risks are unknown.

(b) 100% *inspection*. In terms of quality this method, which consists of inspecting every single item, is undoubtedly the best. Due to fatigue, operator boredom, and distraction, it is not 100% reliable. It is also very expensive.

(c) *Statistical sampling*. This technique requires a sample whose size has been carefully calculated, to be selected at random from the batch. If the sample contains *less* than x defectives, the batch is accepted. If there are *more* than y defectives, the batch is rejected and returned to the supplier for 100% inspection.

Sampling inspection cannot replace 100% inspection in many cases, but where it does the system is considerably less expensive. Furthermore, the risks involved are known and can be allowed for in costing, expected warranty claims, etc.

Such systems are based on:

(a) The sample being representative of the batch.
(b) A knowledge of the probability of acceptance (or rejection) of a batch containing a given percentage of defective items.

It follows that some knowledge of simple probability theory is required.

10.6 SIMPLE PROBABILITY THEORY

Probability is measured on a scale of 0 to 1. An event that is certain to occur has a probability of 1. An event that is certain not to occur has a probability of 0. Thus death and rent day have a probability of 1, and a win on the football pools, having forgotten to post the coupon, has a probability of 0. All other probabilities fall between these extremes.

Further, if an event that occurs can do so in different ways, then the sum of the probabilities of the individual methods of occurrence is 1.

10.61   The Binomial Probability Distribution

If a pack of cards is shuffled and one card drawn from the pack then the probability of it being red or black is equal. The probability of red is 0·5 and of black is 0·5. Let us write down the alternatives. These are:

$$R \text{ or } B$$

and substituting the individual probabilities we get

$$0.5 + 0.5 = 1$$

Doing the same thing if two cards are drawn we get

$$RR \text{ or } RB \text{ or } BR \text{ or } BB$$

Now RB and BR amount to the same thing so we get

$$RR + 2BR + BB$$
$$(0.5 \times 0.5) + (2 \times 0.5 \times 0.5) + 0.5 \times 0.5)$$
$$0.25 + \qquad 0.5 \qquad + 0.25 \qquad = 1$$

∴ The probability of both cards being red is 1 in 4 or 0·25
,,   ,,   ,, both cards being black is 1 in 4 or 0·25
,,   ,,   ,, red and black cards is 1 in 2 or 0·5

Similarly if three cards are drawn the possibilities are:

RRR	RRB	RBR	BRR	BBR	BRB	RBB	BBB
RRR		3RRB			3BBR		BBB

$$0.5^3 \quad + \quad (3 \times 0.5^3) \quad + \quad (3 \times 0.5^3) + 0.5^3 \qquad = 1$$
$$0.125 \quad + \quad 0.375 \quad + \quad 0.375 + 0.125 \qquad = 1$$

∴ The probability of 3 red cards is 0·125 or 1 in 8
,,   ,,   ,, 2 red cards and 1 black card is 0·375 or 3 in 8
,,   ,,   ,, 2 black cards and 1 red card is 0·375 or 3 in 8
,,   ,,   ,, all black cards in 0·125 or 1 in 8

An examination of these three cases

$$R+B$$

$$R^2+2RB+B^2$$

$$R^3+3R^2B+3RB^2+B^3$$

shows that they follow the binomial expansion.

This theorem in fact describes very well the probability of an event occurring in a given way. In fact if an event can occur in two ways, red or black, right or wrong, defective or acceptable, and n events occur, then the successive terms of the expansion are the probabilities of the different methods of occurrence, i.e.

$$(R+B)^n=R^n+nR^{(n-1)}B+\frac{n(n-1)R^{(n-2)}B^2}{2!}+\frac{n(n-1)\ (n-2)R^{(n-3)}B^3}{3!}+\dots$$

If the individual probabilities of R and B, in the case of the cards 0·5 and 0·5, are substituted for R and B in the expansion then the successive terms are the probability of all red; 1 black; 2 black; 3 black cards, etc.

Consider now a bin of components containing $R\%$ good work and $B\%$ bad work. If a sample of n components is taken then the successive terms of the expansion of $(R+B)^n$ give the probabilities of the sample containing 0, 1, 2, 3, etc., defective parts. For instance, if the work contains 10% defectives and 90% acceptable parts, and a sample of 40 is taken we get

$$(0\cdot90+0\cdot10)^{40}=(0\cdot90)^{40}+40.(0\cdot90)^{39}(0\cdot10)+\frac{40\times39(0\cdot90)^{38}(0\cdot10)^2}{2}+\dots$$

$$=0\cdot0148\ +0\cdot065\qquad+\qquad\qquad0\cdot142\dots$$

∴ The probability of 0 defectives in the sample is 0·015

    ,,   ,,   ,, 1 defective ,, ,,   ,, ,, 0·065

    ,,   ,,   ,, 2 defectives ,, ,,   ,, ,, 0·142

Further to this the probability of 1 defective *or less* is the sum of the first two terms $=0\cdot015+0\cdot065$

    $=0\cdot080$

The probability of 2 defectives or less $=0\cdot015+0\cdot065+0\cdot142$

$$=0\cdot222$$

10.62   The Poisson Probability Distribution

Calculation of probability by the binomial system is obviously laborious. Another expansion which relates very well to this work is given by the expansion of $e^x e^{-x}$ in which the successive terms are the probabilities of 0, 1, 2, 3, etc., defectives in the sample, and x is the *expected* number of defectives in the sample.

Now $e^{-x}e^x = e^0 = 1$

and $e^{-x}e^x = e^{-x}\left(1 + x + \dfrac{x^2}{2!} + \dfrac{x^3}{3!}...\text{etc.}\right)$

$$= e^{-x} + xe^{-x} + \dfrac{x^2 e^{-x}}{2!} + \dfrac{x^3 e^{-x}}{3!}...\text{etc.}$$

As tabulated values of e^{-x} are available in normal mathematical tables, log tables, etc., the computation is fairly simple.

Taking the previous case of 10% defectives in the batch and a sample size of 40 we get,

$$x = 10\% \text{ of } 40$$

$$= 4 \text{ defectives expected in a sample of } 40$$

$$\therefore \quad e^{-x}e^x = e^{-4} + 4e^{-4} + \dfrac{4^2 e^{-4}}{2!} + \dfrac{4^3 e^{-4}}{3!}...$$

$$= e^{-4} + 4e^{-4} + \dfrac{16e^{-4}}{2} + \dfrac{64e^{-4}}{6}...$$

$$= 0 \cdot 018 + 0 \cdot 072 + 0 \cdot 144 + 0 \cdot 192...$$

i.e. The probability of 0 defectives in the sample $= 0 \cdot 018$

     ,,   ,,   ,, 1 defective ,, ,,   ,, $= 0 \cdot 072$

     ,,   ,,   ,, 2 defectives ,, ,,   ,, $= 0 \cdot 144$

     ,,   ,,   ,, 3 defectives ,, ,,   ,, $= 0 \cdot 192$

and the probability of 1 defective or less $= 0 \cdot 018 + 0 \cdot 072$

$$= 0 \cdot 090$$

The probability of 2 defectives or less $= 0 \cdot 018 + 0 \cdot 072 + 0 \cdot 144$

$$= 0 \cdot 234$$

and so on.

Examination of these probabilities shows how similar this distribution is to the binomial distribution in section 10.61, the results for the same problem being very similar, but much more easily arrived at.

10.7 CHARACTERISTICS OF SAMPLING SYSTEMS

Assume that a sampling system has been decided upon in which the following instructions are given to the inspector:

'From an incoming batch take a random sample of 40 components.
If the sample contains 1 defective part or less accept the batch.
If the sample contains 2 defectives or more reject the batch.'

This system is denoted as $40_{1/2}$.

It is necessary to know the probability of a batch being accepted (or rejected) if it contains a given percentage of defectives. In section 10.62 it was shown that the probabilities of 0, 1, 2, 3, ... n, defectives are the successive terms of the expression $e^{-x}e^x$.

220

In this case we require the sum of the first two terms for the probability of acceptance, as the allowable number of defectives in the sample of 1 or less (i.e. 1 or 0).

i.e. Probability of acceptance $= e^{-x} + xe^{-x}$

in which x is the expected number of defectives in the sample

$\therefore$ $x = n \times \%$ defectives in the batch

where $n =$ sample size

The best way to show the characteristics of the system is to plot a graph of probability of acceptance against percentage of defectives in the batch. Such a graph, shown for this sampling system in Fig. 10.13, is called the operating characteristic curve.

The points are best found by tabulating the functions in $P = e^{-x} + xe^{-x}$ as shown below.

Percentage Defectives in Batch	$x = Bn$	e^{-x}	xe^{-x}	Probability of Acceptance $P = e^{-x} + xe^{-x}$
0·5	0·2	0·82	0·164	0·984
1·0	0·4	0·67	0·268	0·938
1·5	0·6	0·55	0·33	0·88
2	0·8	0·45	0·36	0·81
3	1·2	0·30	0·36	0·66
4	1·6	0·20	0·32	0·52
5	2·0	0·14	0·28	0·42
6	2·4	0·090	0·216	0·306
7	2·8	0·060	0·168	0·228
8	3·2	0·040	0·128	0·168

Plotting the first column against the last column we get the operating characteristic curve shown in Fig. 10.13.

It should be noted that this operating characteristic curve does not depend on the batch size. In fact a given sampling plan will give the same O.C. curve for any batch size as long as the sample does not exceed about 20% of the batch.

If a larger sample, and acceptance number, is used the plan becomes more discriminating, i.e. a greater number of good batches are accepted and a greater number of bad batches rejected. This alone tends to keep the supplier on his toes. The difference in discrimination is shown in Fig. 10.14 in which O.C. curves are compared for sampling plans for $40_{1/2}$ and $180_{3/4}$ (i.e. sample size $= 180$. Accept the batch if the sample contains 3 or less defectives).

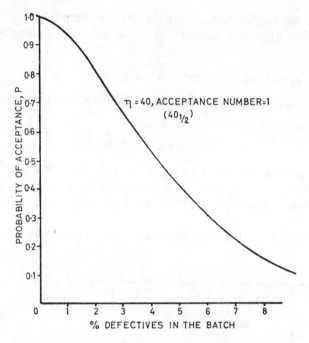

Fig. 10.13.   Operating characteristic curve for sampling system.

The tabulation for $180_{3/4}$ is given below.

Percentage Defectives in Batch	x	e^{-x}	xe^{-x}	$\dfrac{x^2 e^{-x}}{2!}$	$\dfrac{x^3 e^{-x}}{3!}$	P
0·5	0·0	0·407	0·366	0·163	0·049	0·985
1·0	1·8	0·165	0·297	0·267	0·140	0·869
1·5	2·7	0·067	0·181	0·245	0·216	0·709
2·0	3·6	0·027	0·097	0·175	0·209	0·508
3·0	5·4	0·005	0·027	0·073	0·131	0·236

Note $P = e^{-x} + xe^{-x} + \dfrac{x^2 e^{-x}}{2!} + \dfrac{x^3 e^{-x}}{3!}$

Examining these diagrams it is seen that if a batch contains $\frac{1}{2}\%$ defectives, then in both cases it has a probability of 0·98 of being accepted. If a batch has $1\frac{1}{2}\%$ defectives then the sampling plan $40_{1/2}$ gives it a probability of 0·88 of being accepted while the $180_{3/4}$ gives a probability of 0·71 of being accepted and, as the percentage defectives in the batch increases, this divergence in the probability of acceptance grows.

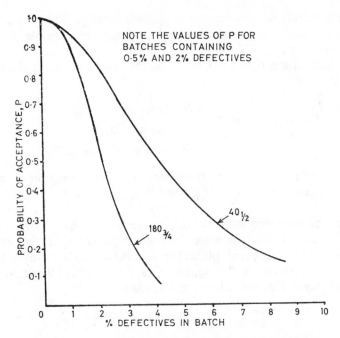

Fig. 10.14. Comparison of O.C. curves for 40½ and 180½ sampling systems.

Two problems now arise. The above two sampling plans were selected arbitrarily but in practice we require the most economic sampling plan. Further, we need to know what percentage defectives are passed into the stores through a receiving inspection department operating a sampling system.

10.8 ECONOMIC SAMPLING PLAN

Any sampling system is fraught with risks. The immediate questions are (*a*) what is the worst batch acceptable? known as lot tolerance percent defective (L.T.P.D.) and (*b*) what risk can be taken of accepting a worse batch than L.T.P.D. due to an optimistic sample? This is known as the consumer's risk.

Any sampling plan chosen must have an O.C. curve passing through the point defined by the consumer's risk on the probability of acceptance axis, and L.T.P.D. on the percentage of defectives in the batch axis. A number of sampling plans will give O.C. curves passing through this point. The most economic plan is the one which gives the required degree of protection for the least total amount of inspection per batch.

In any system all batches have a sample of size n inspected.

All batches rejected have the remainder $(N-n)$ inspected 100%.

Now in the long run a process will produce, when running normally, an

223

average percent defectives, known as the process average. The producer takes a risk that, when running at process average, a pessimistic sample will reject an acceptable batch.

Then the total inspection per batch I in the long run, for a given sampling plan is given by

$$I = n + (N - n)R \text{ where } N = \text{batch size}$$
$$n = \text{sample size}$$
$$R = \text{producer's risk at P.A.}$$

It is seen that I is a function of batch size and R is a function of sample size. If I is plotted against sample size for a given batch size a curve of the type shown in Fig. 10.15 is obtained.

This gives the most economical sample size for this batch size.

This computation is fairly complex and space limits a more full explanation. However, the most economical plans for L.T.P.D. and consumer's risks are published by Messrs. Dodge & Romig, pioneers of this type of work, in their book *Sampling Inspection Tables* published by Wiley, 1959.

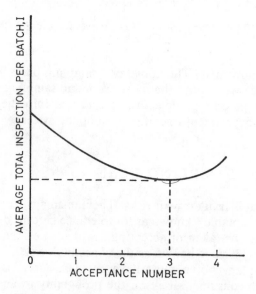

Fig. 10.15. For particular batch size and degree of protection there is a sampling system which requires the least total inspection per batch in the long run. In this case it is the system which allows the batch to be accepted if the sample contains three defectives or less.

10.81 Average Quality Level Passing to Stock

This is known as average outgoing quality level (A.O.Q.L.) and for a given sampling plan can be easily computed from the O.C. curve.

Consider the sampling plan $180_{3/4}$ and assume that a 1000 batches of 1000 parts, all batches containing 1% defectives, have been inspected, using this plan

Now 1000 batches of 1000 gives 1 000 000 components.

From the O.C. curve at 1% defectives 86·9% of the batches will be accepted.

The remaining batches, 13·1% of 1000 will be rejected. Therefore 131 000 parts are rejected.

Now of the 869 000 parts accepted 1% are defective. Therefore defectives in stock are 8690.

Now if the 131 000 rejected items are 100% inspected and each defective part is replaced by a good item, and then returned, the total parts in stock are 1 000 000 of which 8690 are defective.

$$\therefore \text{ Percentage defectives in stock} = \frac{8690}{1\ 000\ 000} \times 100 = 0\cdot869\%$$

Now $0\cdot869\%$ = Percentage defectives in batch $\times$ Probability of acceptance

$\therefore$ A.O.Q.L. = Percentage defectives in batch $\times$ Probability of acceptance

Tabulating the results for the $180_{3/4}$ plan we get

Percentage Defectives in Batch	Probability of Acceptance	A.O.Q.L.
0·5	0·985	0·493
1·0	0·869	0·869
1·5	0·709	1·064
2·0	0·508	1·0160
3·0	0·236	0·708

Plotting A.O.Q.L. against percentage defectives in batch, a graph as in Fig. 10.16 is produced.

Thus the worst possible A.O.Q.L. is likely to be approximately 1·1% and it is unlikely that this will be reached. The limit will only occur if all batches come in at about 1·7% defectives, which is unlikely. As soon as batches arrive with a greater or lower percentage defectives, the percentage defectives passing to stock is reduced. The good batches have few defectives and the system ensures that the majority of the bad batches receive greater attention.

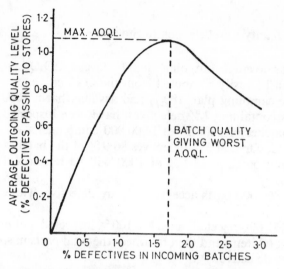

Fig. 10.16.   Graph of A.O.Q.L. plotted against
percentage of defectives in the batches.

10.82   Double Sampling

The single sampling system described in the previous section greatly improves the economics of inspection. However, it must be realized that inspection of any item is usually more complex than looking to see if it is the right colour. Complete inspection of a sample of 180 units represents a lot of work.

Now if a batch is very good or very bad a small sample will detect it. This was recognized by Dodge and Romig who devised a system of double sampling to reduce the total inspection. In this system the inspection instructions are as follows:

(*a*) From a batch select a random sample n_1.

(*b*) If the sample contains x defectives or less accept the batch.

(*c*) If the sample contains y defectives or more reject the batch.

(*d*) If the sample contains more than x but less than y defectives select a further sample of n_2 components.

(*e*) If the total sample ($n_1 + n_2$) contains less than y defectives accept the batch.

(*f*) If the total sample contains y defectives or more, reject the batch.

The Dodge–Romig tables also contain details of this double sampling system, and should be consulted for further details of this type of work.

10.9 CONCLUSION

Many books, some extremely lengthy, have been written on the subject of statistical quality control. This chapter can only condense the essential ingredients into what is hoped is a digestible form. It is intended to stimulate interest in this subject which, like those discussed in other chapters, is a tool of the inspector and metrologist, the importance and complexity of which is constantly growing.

It is further emphasized that the application of statistics is not the cure to all industrial ailments; but the procedures, and their results, suggested in this chapter generally hold good in the long run. One sample does not give detailed information about one batch, but over a period of time a series of samples gives a great deal of information about a lot of batches.

APPENDIX TO CHAPTER 6

Determination of the Flatness of a Plane Surface

IN Chapter 6 it is demonstrated how an auto-collimator, or a spirit level, may be used to measure the deviation from straightness of a machine tool guide-way. The same principles may be used to determine the deviation from a true plane of a large surface such as a surface table or machine table.

A flat surface is composed of an infinitely large number of lines, or generators, and for it to be truly flat the following conditions must be satisfied:

(*a*) All generators must be straight.

(*b*) All generators must lie in the same plane.

It should be noted that provided condition (*a*) is *completely* realized then condition (*b*) must also hold good. The two conditions are emphasized as it is the verification of condition (*b*) which is the main problem. Also it must be realized that it is not a sufficient test, in the case of a rectangular surface, to measure the straightness of generators parallel to the edges. These may all be straight but the surface need not be flat.

Consider a sheet metal box having a pair of diagonally opposite corners reduced in height, but whose sides are straight. If the box is filled with plaster of paris which is then levelled off with a straight edge which is kept parallel to one end, then all lines across the surface must be straight (they were produced by a straight edge). Similarly all lines at 90° to these generators must be straight, as the straight edge was controlled by two other straight lines, these being the edges of the box. Thus if such a surface is tested for flatness along lines parallel to its sides it will appear to be flat. That it is not is clearly seen from Fig. 11.1, it being concave across one diagonal and convex across another.

It is immediately seen that if the surface is to be verified as being truly flat then it is necessary to measure the straightness of the diagonals, in addition to the generators parallel to the sides.

The measurement of straightness of all of these lines of test may be carried out with an auto-collimator as is described in Chapter 6, but having made these measurements it is necessary to relate each line of test to all of the others, i.e. verifying conditions (*b*) with which this appendix is concerned.

Consider the surface shown in plan view of Fig. 11.2 on which the eight main generators are set out. These should be chosen just inside the edges of the table so that the edge area, which is prone to damage, is avoided. The length of the lines should be whole multiples of the length of the base of the spirit level or reflector stand, whichever instrument is used, and it is advisable to select side and diagonal lengths in the ratio of $3 : 4 : 5$.

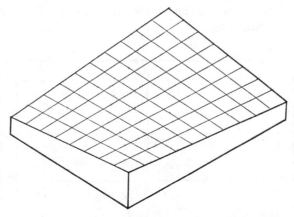

Fig. 11.1. A surface, all of whose generators parallel to the sides are straight, but which is not flat.

The procedure is as follows:

(*a*) Carry out a normal straightness test on each generator.

(*b*) Tabulate each set of results only as far as the cumulative error column.

(*c*) Correct the ends of AC; AG; and CG; to zero. This gives the heights of points A, C, and G as zero and these three points then constitute an arbitrary plane relative to which the heights of all other points may be determined.

(*d*) From (3) the height of O is known relative to the arbitrary plane ACG=OOO. As O is the common mid-point of AE, CG, BF, and HD, all points on AE are now fixed. This is done by leaving A=O and correcting O on AE to coincide with the mid-point O on CG.

(*e*) Correct all other points on AE by amounts proportionate to the movement of its mid-point. Note that as E. is twice as far from A as the mid-point, its correction is double that of O, the mid-point.

(*f*) As E is now fixed and C and G are set at zero, it is possible to put in CE and GE, proportionally correcting all intermediate points on these generators.

(*g*) The positions of H and D, and B and F, are known so it is now possible to fit in lines HD and BF. This provides a check on previous evaluation since the mid-point of these lines should coincide with the known position of O, the mid-point of the surface.

Thus the height of all points on the surface are known, relative to an arbitrary plane ACG; but this may not be the best plane and correction must be made for this.

However, consider now an example illustrating the method outlined to relate a series of test lines to each other.

The table below is a set of *cumulative* errors for the lines of test designated in Fig. 11.2 on a surface table.

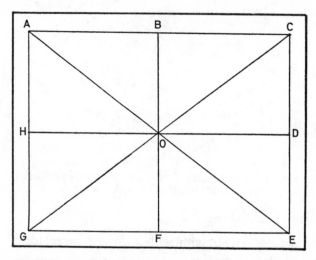

Fig. 11.2. Surface table marked out with the minimum number of lines for a flatness test.

Cumulative Errors of Individual Lines of Test

A–C	*A–E*	*A–G*	*C–C*	*G–E*	*C–E*	*B–F*	*H–D*
0	0	0	0	0	0	0	0
0	0	0	0	0	0	0	0
−1	0	+1	+2	+1	−1	+1	+3
−4	−1	+2	+4	−3	+2	+2	+7
−7	−2	−2	+5	−6	+5	−2	+9
−12	−4	−6	+6	−8	+3	−5	+9
−15	−8	−6	+4	−9	+2	−7	+6
−15	−12		+2	−11			+9
−18	−17		0	−12			+10
	−21		−2				
	−24		0				

It is convenient now to consider these lines of test on a plan view of the surface as in Fig. 11.3 in which lines AC, AG and CG have been corrected to zero at each end. Thus the plane ACG is fixed with the points A, C, and G at zero, and points on these three lines are all known relative to this plane.

It is seen that the mid-point is positioned at +6 units above the plane, and the mid-point of line AE must coincide with this position, while point A is known to be O.

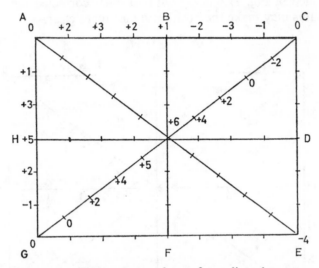

Fig. 11.3. Three corners of a surface adjusted to zero enable the height of the mid-point to be fixed relative to a plane through the corners. This enables the height of the other corner to be determined.

Correction for Line A E

Cumulative Error	Correction	Height Relative to Plane ACG
0	0	0
0	+2	+2
0	+4	+4
−1	+6	+5
−2	+8	+6
−4	+10	+6
−8	+12	+4
−12	+14	+2
−17	+16	−1
−21	+18	−3
−24	+20	−4

231

From the table of cumulative errors the value of the mid-point of AE is seen to have a value of -4 units. For this to become $+6$ units it must be raised by $+10$ units and thus point E, which is twice as far from A, must be raised by $+20$ units, giving E a final value of $(-24+20) = -4$ units. All other points on AE are corrected by proportionate amounts, so that a table for AE may be drawn up as shown below.

These values may be inserted on the diagram of the surface as in Fig. 11.3. They are included in Fig. 11.4 along with all other corrected figures, as the two separate diagrams may make the position rather more clear.

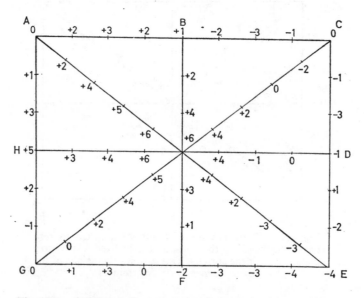

Fig. 11.4.   Height of all measured points related to an arbitrary plane ACG.

The height of point E being known as -4 units, relative to plane ACG, enables the relative heights of all points on lines CE and GE to be fixed relative to this plane. Considering line CE it is seen the value of E in the table of cumulative errors is $+2$ units. Hence to make it -4 units it must be corrected by the amount -6 units and all other points corrected by proportional amounts.

Similarly on line GE, point E has a value of -12 units in the table of cumulative errors and it must therefore be corrected by $+8$ units, and by proportional amounts on intermediate points. These tables of corrected values are shown below.

It remains now only to fix all points on lines BF and HD relative to the plane ACG.

Considering line BF it is seen that relative to plane ACG, point B has a value of $+1$ unit, but the value of point B in the table of cumulative errors is O, so that initially all points on BF must be increased by $+1$ unit.

Correction for Line C E

Cumulative Error	Correction Rel. to ACG	Error Rel. to ACG
0	0	0
0	−1	−1
−1	−2	−3
+2	−3	−1
+5	−4	+1
+3	−5	−2
+2	−6	−4

Correction for Line G E

Cumulative Error	Correction Rel. to ACG	Error Rel. to ACG
0	0	0
0	+1	+1
+1	+2	+3
−3	+3	0
−6	+4	−2
−8	+5	−3
−9	+6	−3
−11	+7	−4
−12	+8	−4

Then point F, whose cumulative value, corrected by +1 unit, becomes −8 units, must be made to coincide with its known value relative to plane ACG of −2 units, i.e. its value must be increased by +6 units and intermediate values corrected by a proportional amount.

If a similar process is applied to line HD, as in the tables below, then the values of the points relative to plane ACG may be inserted in Fig. 11.4.

Correction for Line B F

Cumulative Error	Initial Correction	Correction	Error Rel. to ACG
0	+1	0	+1
0	+1	+1	+2
+1	+2	+2	+4
+2	+3	+3	+6
−2	−1	+4	+3
−5	−4	+5	+1
−9	−8	+6	−2

Correction for Line H D

Cumulative Error	Initial Correction	Correction	Error Rel. to ACG
0	+5	0	+5
0	+5	−2	+3
+3	+8	−4	+4
+7	+12	−6	+6
+9	+14	−8	+6
+9	+14	−10	+4
+6	+11	−12	−1
+9	+14	−14	0
+10	+15	−16	−1

It should be noted that the mid-points of both of these lines of test coincide correctly with the value of +6 units for the mid-point of the surface. This provides a useful check on the calculations up to this point.

It may be thought that this is the end of the matter, but this is not so, because the plane ACG was chosen entirely arbitrarily, and the definition of flatness error states that the departure from flatness is the *minimum* separation of a pair of parallel planes which will just contain all points on the surface. Consider a surface as shown in Fig. 11.5 (*a*) in which three corners have heights of zero, relative to same arbitrary plane, and the fourth corner has a value of +10 units relative to this plane. It might be thought that the departure from flatness is +10 units, but if the plane is allowed to tilt about the axis XX and the two opposed free corners allowed to become equal as in Fig. 11.5 (*b*) it is seen that the departure from flatness is only +5 units.

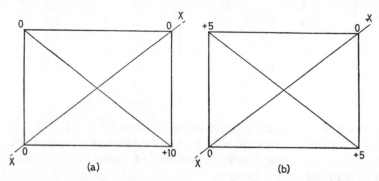

Fig. 11.5(*a*). Initial assessment shows a flatness error of +10 units at one corner relative to an arbitrary plane. (*b*). By tilting the whole surface about axis XX, the actual error is shown to be +5 units.

If this procedure is to be followed for the surface shown in Fig. 11.4, it is seen that the amount any given point is raised or lowered, depends on its distance from the axis. Thus the calculation for this final correction to determine the minimum separation of a pair of parallel planes which will just contain the surface, can become extremely laborious, particularly when it is realized that the process must be carried out at least twice, on axes at right angles to each other.

A possible simplification of this process has been suggested, using a graphical method as outlined below. If we consider again Fig. 11.5 (*a*) and make a projection of the surface along the line of tilt we see the surface as in Fig. 11.6.

It is seen that a pair of parallel lines may be drawn, which just enclose all points on the surface, whose separation is much less than +10 units. In fact, if the scale is considered, it is 5 units as was found by tilting.

To apply this technique to the points on a surface such as that in Fig. 11.4 the procedure is as follows.

(*a*) Arrive at the condition shown in Fig. 11.4 and select two points, preferably on opposite sides, whose values are the maximum positive and maximum negative relative to the arbitrary plane, in this case ACG. Connect these points and project at right angles to the line XX connecting them.

(*b*) Set off to scale the height of all points relative to a line YY, parallel to XX, which represents plane ACG.

234

(*c*) By inspection select the closest pair of parallel lines which will contain all of the points. It should be noted that one line will have two points on it, and the other line one point.

(*d*) Draw a centre line ZZ between these two and refer all points to this line.

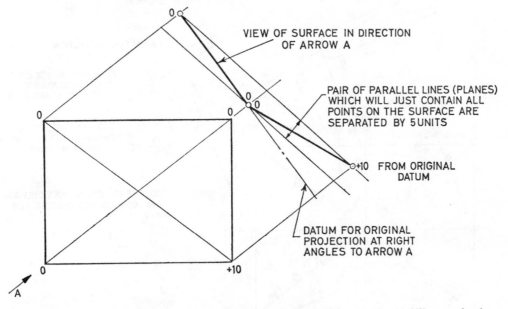

Fig. 11.6.   The true flatness error of +5 units, obtained by tilting in Fig. 11.5(*b*), can also be obtained by projection.

It is important to realize that the two parallel lines represent planes at right angles to the plane of the paper. It may be possible to bring them still closer by inclining them, as a pair, to one side or the other. This can be done by repeating the above process, i.e. draw another plan view of the surface inserting the results from (*d*) above, and project again at right angles to the line of the original projection.

This procedure has been carried out for the surface referred to previously, the results being shown in Fig. 11.*l*.

It must be emphasized that this is not an exact method. It contains an error due to the differences in scales for lengths and heights of the surface. Also more than two projections may be required but in practice it has been found that the percentage reduction in the separation of the parallel planes containing the surface, by continued projection, is not significant unless the line of the original projection is particularly badly chosen.

Another method of carrying out this process is to refer all points to x, y and z axes, thus fixing them in space. It is then possible to determine the minimum separation of the parallel planes containing the surface by finding the best plane

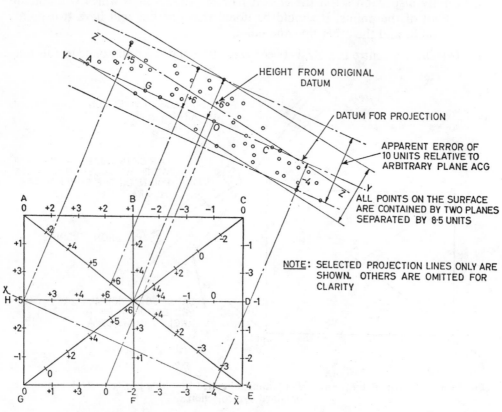

Fig. 11.7. Determination of flatness error by graphical methods.

The first projection only is shown. It may be necessary to refer all points to line ZZ and project again at 90° to the original projection, but in this case it is unlikely as some of the new maxima are widely spaced.

so that the sum of the squares of all points from it is a minimum. This is an extension of the method of least squares (1·51) operating in three dimensions. With a large number of points to be considered, a computer is necessary for this calculation.

Finally it must be realized that whatever method is used it is a laborious process. Many more points would be taken than have been used in the example, a complete grid of the surface being tested, all lines related, and cross-checked in the calculations. The authors feel that without the aid of a computer the effort involved by the graphical method increases least as the number of points surveyed increases.

Bibliography

BARNARD, C. P., 'Notes on Circular Dividing Apparatus' (N.P.L.), H.M.S.O.

BARRELL, H., 'The Bases of Measurement', Sir Alfred Herbert Paper, 1957, I. Prod. E.

—— 'Engineering Dimensional Metrology' (N.P.L. Symposium), Vols. I and II, H.M.S.O.

CHAPMAN, W. A. J., *Workshop Technology*, Parts II and III, Arnold.

EVANS, J. C., 'The Pneumatic Gauging Technique in its Application to Dimensional Measurement', J. I. Prod. E., **36**, No. 2, 1957.

HABELL, K. J. and COX, A., *Engineering Optics*, Pitman.

HEMSLEY, S. H., *Optical Instruments in Engineering*, Paul Elek.

HUME, K. J., *Engineering Metrology*, MacDonald.

—— 'Developments in Dimensional Accuracy', J. I. Prod. E., **40**, No. 5, 1961.

JUDGE, A. W., *Engineering Precision Measurements*, Chapman & Hall.

KING, G. K. and BUTLER, C. T., *Principles of Engineering Inspection*, Cleaver-Hume.

LOXHAM, J., 'The Control of Quality in Engineering Manufacture', Proc. of Conference on Technology of Engineering Manufacture, I. Mech. E., 1958.

—— 'An Experiment in the use of a Standard Limit System', Proc. I. Mech. E., **156**, 1947.

—— 'The Potentialities of Accurate Measurement and Automatic Control in Production Engineering', J. I. Prod. E., **39**, No. 12, 1960.

MERRITT, H. E., *Gears*, Pitman.

NICKOLS, L. W., 'The Effective Diameter of a Parallel Screw Thread', J. I. Prod. E., **40**, No. 5, 1961.

PARKINSON, A. C., *Gears, Gear Production and Measurement*, Pitman.

POLLARD, A. F. C., *Kinematical Design of Couplings*, Adam Hilger.

—— 'Kinematic Design in Engineering', Prod. I. Mech. E., **125**, 1933.

ROLT, F. H., *Gauges and Fine Measurements*, Vols. I and II., Macmillan.

—— 'The Development of Engineering Metrology', Sir Alfred Herbert Paper, I. Prod. E., 1952.

SCHLESINGER, G., *Testing Machine Tools*, Machinery Pub. Co.

SHARP, K. W. B., *Practical Engineering Metrology*, Pitman.

—— 'Notes on Inspection Organisation and Dimensional Control' (Admiralty), H.M.S.O.

—— 'Gauge Making and Measuring' (N.P.L.), H.M.S.O.

—— 'Gauging and Measuring Screw Threads' (N.P.L.), H.M.S.O.

STEEDS, W., *Mechanism and Kinematics of Machines*, Longmans, Green.

TANNER, C. J., 'Air-Gauging—History and Future Development', J. I. Prod. E., **37**, No. 7, 1958.

TIMMS, C., 'Recent Development in Spur and Helical Gears', J. I. Prod. E., **39**, No. 6, 1960.

TOWN, H. C., and COLEBOURNE, R., *Engineering Inspection, Measurement, and Testing*, Odhams.

TWYMAN, F., *Prism and Lens Making*, Adam Hilger.

Index